THE
# BED BUG
# SURVIVAL
# GUIDE

THE

# BED BUG SURVIVAL GUIDE

## The Only Book You Need to Eliminate or Avoid This Pest Now

### JEFF EISENBERG

**GRAND CENTRAL**
PUBLISHING

NEW YORK    BOSTON

Grand Central Publishing
Hachette Book Group
237 Park Avenue
New York, NY 10017
www.HachetteBookGroup.com

Printed in the United States of America

First Edition: April 2011
10 9 8 7 6 5 4 3 2 1

Grand Central Publishing is a division of Hachette Book Group, Inc. The Grand Central Publishing name and logo is a trademark of Hachette Book Group, Inc.

The publisher is not responsible for websites (or their content) that are not owned by the publisher.

Library of Congress Cataloging-in-Publication Data
Eisenberg, Jeff
   The bed bug survival guide : the only book you need to eliminate or avoid this pest now / by Jeff Eisenberg.
      p. cm.
   ISBN 978-0-446-58515-6
   1. Bed bugs—Control.   I. Title.
   TX325.E45 2011
   648'.7—dc22                                    2010049217

*Dedicated to the four pillars of my life*

*Shira Eisenberg*

*Mom*

*Yussi and Miriam Lobel*

———◆———

*100% of the royalties I receive from sales of this book, excluding agent fees, will be donated to the Tikun Olum Foundation and American Friends of Migdal OHR.*

# Acknowledgments

I want to thank my wife, Shira, for providing me the necessary time to write this book. Her patience, confidence, and support were integral in helping create this timely book. I want to apologize to my four amazing children, Tamar, Zevi, Shayna, and Shmaya, who did not always get my full attention when they may have needed it, but who showed me such tremendous patience, understanding, and love.

I want to express my warmest thanks and appreciation to Ayana Byrd, Kenrya Rankin, and Richard Simon for helping make this book possible. Their enthusiasm and sincerity in helping people better cope with this epidemic served as an ongoing inspiration. I also want to thank Mary Evans, one of my two agents. She has a sterling reputation and is the paradigm of elegance and class. I was honored by her acceptance to represent me on this project. To my second agent, Tanya McKinnon, I extend my infinite gratitude. I had never considered writing a book on bed bugs until she knocked on my office door that beautiful spring day. Her persistence and absolute confidence in me to help bed bug sufferers everywhere was compelling. Her magnanimous offer to refuse any fees for her efforts is a testament to her dedication to the underlying purpose of this book.

My sincerest gratitude to my editor at Grand Central Publishing, Diana Baroni, who not only believed in the project and shaped the book, but is now close to an expert herself in the prevention of bed bugs!

I want to thank my wonderful and patient office staff—Shari Clarke, David Robinson, Sherrell Hackett, and Claire Villard—who calm the nerves of tens of thousands of frantic customers with such warmth and sympathy. They are constantly learning and adapting to the many changing bed bug protocols at my company and graciously put up with all the different material and information I throw at them daily.

I want to thank Eric Baum from Simon, Eisenberg & Baum, one of the most respected real estate attorneys in New York City; Wayne Tusa from Environmental Risk and Loss Control, Inc., a rare and genuine environmental consultant; Larry Pinto of Pinto and Associates; Bob Rosenberg at the National Pest Management Association; Dr. Changlu Wang from Rutgers University; Dr. Phil Koehler at the University of Florida; and Dr. Josh Benoit at Yale.

We make a living for what we get,
we make a life for what we give.

—WINSTON CHURCHILL

# Contents

# Contents

## Part Two: Treatment

## Part Three: The Past and Future

THE
# BED BUG
# SURVIVAL
# GUIDE

# Introduction

## Bed Bugs Happen to *Other* People, Right?

Right. They happen to other people such as:

- the Hollywood costume designer who got them from a beautiful couch purchased at a reputable Beverly Hills vintage store
- the five-year-old who carried them home in his *Yo Gabba Gabba!* backpack, which he grabbed from the communal pile in the school coatroom
- the young administrative assistant who got them from her new work cubicle
- the Condé Nast fashion editor who doesn't know where they came from, but does know she spent $5,000 and threw away irreplaceable treasures—including her Vera Wang wedding gown—to get rid of them

So, yes, bed bugs always happen to other people. Until they happen to you.

As of right now, there are no more than three

degrees of separation between you and someone who has had these tiny vampires in their home.

In 1996, I had my first run-in with bed bugs. A hostel called my company because visitors were complaining about being bitten in the night. It took hours of turning over mattresses and looking behind bureaus until I finally found what I thought was the culprit. After working with an entomologist to figure out what the six-legged, quarter-inch-long insect was (virtually no exterminator had seen them or dealt with them since the '60s), I was faced with an even bigger challenge: how to kill the darn thing. Because we thought they'd all died out, there was no established protocol for tackling them, and the favored methods of the old-timers I did manage to track down were either illegal or outdated. What began with an unidentifiable bug in a New York City hostel quickly became the focus of my work life and has been for the past fourteen years.

Over the next several years I studied bed bugs, experimenting with different techniques through a process of trial and error. I kept meticulous records of my findings, and was continuously fascinated by the tenacity and evasiveness of this flat little bug. As a result, my program for dealing with it has been evolving every three months since my first encounter with them. I was able to develop an effective system for eliminating infestations. In the past few years I've been able to make that process greener by including heat and other nonchemical treatments, but that was only *after* teaching clients how to prep their spaces for maximum success. And over that time the calls poured in from co-ops, condos, businesses, colleges, hotels, schools, landlords, homeowners, tenants, and other exterminators as they realized I was one of the only experts in New York City with an evolving bed bug protocol that worked.

Still, when I rang the alarm to my peers at an industry convention in subsequent years, saying that things were quickly getting out of control, no one paid much attention. Then in 2005 the *New York Times* ran a page-one piece declaring that bed bugs were back. Finally, the word was getting out. But even in 2006 when the New York City Council asked me to testify about the spiraling numbers of bed bug cases, not even the mayor paid attention to my warnings that we had a burgeoning epidemic on our hands.

I hadn't set out to be the bearer of bad news, but I felt I had no choice. One conservative estimate put sightings up more than 70 percent nationwide since 2007. Reported infestations in Chicago tripled between 2007 and 2008. Based on the rapidly increasing volume of business my company and other companies operating within the same service area have been doing, I project that by the time you're reading this at least 85 percent of all New York City buildings will have had or have a problem with bed bugs. I've personally seen bed bug cases double each year since 2006—they now make up 40 percent of my business. And it's not just a "city problem" anymore; suburbs and rural towns all over the world from Fox Chase, Kentucky, to Singur, India, are doing battle. As far as bed bugs are concerned, we've reached the tipping point.

However, as I always tell my clients, while the big numbers are bad, the worst part is on a human scale—the way sufferers are stripped of their normal lives. Sometimes it takes months for people to realize they have bed bugs. By the time they've wrapped their brains around reality, they have a full-scale infestation. Then they feel scared and helpless, and often ashamed. Without reliable information or help, they are prey to a growing breed of charlatans who capitalize on their fear and ignorance, needing repeated exterminations because of mishandled jobs. Single people stop dating for fear of passing the bugs along. Parents cancel play dates because they don't want to risk their child carrying the pest into someone else's home on a favorite toy or teddy bear; other parents who have heard the news start keeping *their* kids away. Neighbors become hostile, landlords turn a deaf ear, and you can find yourself very isolated, sitting home alone and reading terrifying stories online.

I regularly give interviews for newspapers, magazines, radio, and TV news shows. I've won several awards for my focus on environmentally friendly treatment methods. I provide expert testimony during legal proceedings around the country, and I help entire industries develop written protocols for handling bed bug infestations. I receive no fewer than three hundred local calls and e-mails for help each week. And then there are the international SOS calls—at least twenty a week from Europe alone. Yet the people who call me from across town or across the ocean aren't looking for an exterminator; they've already had exterminators, and it wasn't enough. They call me because they need a strategist.

And that's what I am, first and foremost—a strategist in identifying the source, treating infestation, and, above all, preventing future issues. I liken it to seeing a doctor who cares. If you went in complaining of severe headaches and dizziness, you wouldn't be satisfied with "Take two aspirin and call me in the morning." You'd want your doctor to pepper you with questions and run a battery of tests to figure out what's wrong and then heal you.

I am your bed bug doctor. I ask a million questions, from "Do you live alone?" "Where do you work?" "Do you travel often?" "Do you have a nanny?" to "Have you seen any mattresses outside your place lately?" to "Are your kids visiting from college?" (Sounds nosy, but intrusive questions are the only way to track down the source of an infestation, and they have helped me solve thousands of cases. I can treat your home a million times, but if your boyfriend is bringing a fresh supply of bed bugs to every sleepover, you will never be rid of them.) I check out every possibility. I know what to use and when to use it, always taking into account your lifestyle, health issues, family, pets, the type of home you live in, and the severity of the infestation.

In other words, I'm not just some guy with a spray

can, I'm as much a psychologist, an advocate, a scientist, an environmentalist, and an empathic best friend as I am an exterminator. I bridge the gap between lab-coated entomologists and what you need to know *right now*. By reading a book that lays out my fourteen years of experience treating bed bugs you can get the best of my help—exact information on how to check for bed bugs, reduce clutter, decontaminate your belongings, and pack so that extermination can even begin. I also guarantee that following my prevention methods will cut the likelihood of your getting bed bugs by at least 75 percent. And if you do get them, my advice will save you at least $1,500, not to mention months in false starts and unqualified exterminators as you try to get rid of the bugs. Once you are rid of them, this book will help *keep* you rid of them.

## How Can This Book Cure an Epidemic?

Few animals are without parasites. Sharks have pilot fish; deer have ticks; horses, of course, have horse-flies. And until about fifty years ago, humans had bed bugs. These creatures feasted on pharaohs and presidents, kings and queens, and everyone else. I'm not saying we should embrace our long-lost parasite and welcome it back to the fold. But understanding that this is a historical problem, with only a half-century respite since man stood upright, underlines two crucial points in the bed bug battle:

1. People's shame helps bed bugs thrive; and
2. Eradicating bed bugs requires you to be an active, *not a passive*, participant. Picture yourself playing football in the mud: if you hop in the shower in your uniform and cleats you can wash off *some* of the muck, but you won't really get clean. The same goes for bed bugs. You cannot get rid of them all without a *serious* plan that requires more than simply handing a check to an exterminator and hoping he knows what he's doing. You are a key component of your own bed bug eradication. You must work with a pest management professional (PMP) before he comes and after he leaves, to insure that you are really rid of them and that you *stay* rid of them.

If you opened this book to find the most effective ways to get rid of bed bugs, you came to the right place. That's what I help folks do every day—well over 150,000 people so far. *But my biggest goal is to keep you from getting them in the first place.* That's the key to stopping the epidemic and keeping you more or less sane (at least when it comes to bed bugs).

Eradicating bed bugs does not hinge on traditional exterminating practices but on prevention—and to do that, you need to change some very basic things about the way you travel, clean, work, shop, sleep, and socialize with friends. It takes a holistic approach that looks at your entire living situation—and that's the approach I use each time I successfully tackle someone's infestation or help them avoid one in the first place.

I won't even treat a space or home where they're not taking education and prevention as the most serious component—as I always say, "Getting rid of bed bugs takes teamwork!" Almost every client who contacts me has the same three questions:

1. Where do bed bugs come from?
2. How do I know if I have them?
3. How do I get rid of them?

The answers to those questions alone are (literally) enough to fill a book, yet they still don't get at the real meat of the matter. The question they *should* be asking is: "How can I prevent a bed bug infestation?" The answer is *education, education, education*. That's the only shot we have at shutting down this epidemic, and although it is an intense process, this book will walk you through it step-by-step in a way that I have proven will work. This book will answer *all* these questions so that we can stop giving bed bugs to each other at home, work, and everywhere else we go.

In this book I employ a clear, highly usable format to put you on the most immediate, effective course of action—one that I spent the last decade-plus figuring out and perfecting, so you don't have to waste your time with what *doesn't* work. As I said before, following my prevention methods will reduce your chances of getting bed bugs by 75 percent and save you at least $1,500 in missteps while ridding your life of the problem if you do get them.

## How to Use This Book

Think of *The Bed Bug Survival Guide* as a BFF for the freaked out. Whether you think you're being snacked on, want to prevent an infestation, or are just curious as hell, it's there for you. It will help you track down any unwanted houseguests, rub your back while you cry a little, hold your hand while you find an exterminator, save you from spending thousands of dollars on "fixes" that are worthless (and potentially dangerous to you, your children, and your pets), and do everything possible to keep you from going through it again. And like your BFF, it'll be honest and to the point, with short snippets of info that say exactly what you need to know to feel confident, but not so much that you feel overwhelmed—no long, boring speeches here. You can read it straight through, or go directly to the info that most concerns you right now. So if you're heading to the movies, you can brush up on how to prevent picking up bed bugs with your ten-dollar extra-small popcorn.

I've designed this book so it's easy to use. If you think you may have bed bugs read chapter 1 to learn how to inspect your home for them, and if you don't have them read chapter 2 to learn how to prevent getting them. If you're sure you have bed bugs read chapter 2 followed by chapters 6 and 7. Regardless of whether you have them or not, chapter 1 cannot be skipped. Prevention is the key to avoiding them *as well as* getting rid of them. As you will read, bed bugs are not like other pests. Keeping your home free of them requires ongoing awareness, vigilance, and specific cleaning techniques.

There is no quick fix or silver bullet for beating bed bugs—but there *is* real truth, there *are* real tools, and there *are* proven techniques for fighting this minuscule menace. And it is all right here, in *The Bed Bug Survival Guide*, which is the first comprehensive handbook for the layman that cuts through the noise of urban legends and misinformation to empower people with practical strategies for preventing and treating this growing problem.

So whether you're a mom who's freaking out that her daughter's sleepover guests will leave more behind than a few stray socks, a college student who just moved into a dorm and now can't stop scratching, or a businessman who stays in a different hotel every week and isn't particularly fond of bloodsucking hitchhikers, this book will be your most helpful tool in protecting your home and family from bed bugs. And if they happen to sneak past your vigilant watch, it will guide you through a physically and emotionally difficult process intact. It will save you time, money, mental anguish, anxiety, lost sleep, family disorder, physical distress, and social ostracism.

One of my clients, a media mogul, said, "I wouldn't

wish this on my worst enemy"—and she had a lot of enemies! Her words speak powerfully to how devastating it can be to have bed bugs. The epidemic is as bad as I'd ever imagined it could be, and it's getting worse. *But wait*—now is not the time to assume the fetal position in (someone else's) bed and deny that we have a problem. Now is the time to arm ourselves with effective info and use it to help stop, or at a minimum slow down, this epidemic once and for all.

Good night and sleep tight!

# 1

## First Things First

What Are Bed Bugs and Do You Have Them?

Once upon a time, though this is no fairy tale, I got a phone call from a woman living in New York City. She frantically told me how she had never even heard of bed bugs outside of a nursery rhyme until suddenly the newspapers, the nightly TV news, and the Internet were full of stories about them. So as soon as she suspected there was a problem, she sprang into action and called the landlord. In turn, the landlord called the building's regular exterminator, who treated the problem like it was a roach job. The poor woman on the phone explained to me that she thought everything was fine once the exterminators left—until she had new bites a few weeks later. And so did her nanny. And her daughter's best friend, who had come for a sleepover. A month later her coworkers did, too. She thought she had solved the problem, but instead it was getting worse. And so, months after the problem first began, she called me.

Here's the happier ending this woman, like so many that I talk to every day, *could* have had: a few hundred dollars spent to confirm there was a problem, followed by a pest management professional that would have cost less than $1,500, all wrapped up in a matter of weeks.

What's the lesson here? When it comes to bed bugs, you and your pocketbook cannot afford to play the "I don't want to think about it" card or the "I'm just going to trust someone else to handle it" one. They may be "disgusting" and "gross" and make you uncomfortable to consider, but refusing to face up to bed bugs and not seeking out an expert will cost you money, time, and a lot of sleep. The good news: you *are* in some control—a good deal, actually. So instead of popping Valium or burying your head in the sand, fight back!

Your best line of defense? Know thy enemy! Bed bugs are nothing new. They're the same creatures that plagued the ancient Egyptians, the royal family of England, and former residents of 1600 Pennsylvania Avenue—so you're actually in good company. With that in mind, we answer a few of your Top Bed Bug 101 questions here:

## What the #@!% *Is* a Bed Bug?

The scientific name for bed bugs, *Cimex lectularius*, is Latin for "lounging bug" or "bug of the bed." More facts:

- A bed bug is a wingless insect, a parasite that likes to rest during the day and come out when its food source is at rest or motionless (which is usually, but not always, at night).

- They are not known to spread disease.
- Bed bugs have six life stages and must go through five nymphal stages before becoming an adult (and capable of reproduction). In order for bed bugs to get from one stage to the next they must feed on blood (yours, your pets', or other mammals'). They are almost impossible to see in the first few stages without proper magnification (the tiny translucent or white eggs grow to be flat, oval, and with a short head) until they reach adulthood, when they grow to be about the size of an apple seed and turn a reddish brown (because they're now filled with your blood).
- The female can lay up to four eggs a day, in secluded locations (like the zipper of your handbag or your briefcase). Typically she can lay five to ten a week and up to five hundred in her lifetime.
- They can fit into any crack the width of a business card.
- The tip of the female's abdomen is rounded and that of the male's abdomen is pointed. Though you probably don't ever want to get close enough to tell, this is useful information to an exterminator.
- They're attracted to carbon dioxide, which we breathe out, as well as to our body heat.
- They have been known to live up to eighteen months without a blood meal, under ideal conditions. Typically it is not more than ten months.

- They typically spend five to ten minutes at a time feeding.
- They prefer to eat in the dark (more privacy, less chance of you being awake and moving), but they will contentedly chow down during the day if that's when their host is available.
- The hungrier they are, the bolder they will be in their hunt for a meal. Luckily, their recklessness increases the chance of you spotting them as they come into more visible places.
- They live for up to one and a half years, but sometimes longer.

## Quiz: How Bed Bug Proof Are *You*?

When it comes to bed bugs, you are only as safe as your lifestyle. Before I start with the checklists and dos and don'ts, let's take a look at your current habits with a little quiz. For better or worse, it'll help you get a handle on just how safe you really are, and give you something on which to build.

1. You're having dinner with friends. You sit down and put your bag
   a. in your lap.
   b. on the back of your chair.
   c. on the floor near your feet.

2. You have houseguests
   a. maybe once a year.
   b. once a season or so.

c. there's a friend of a friend here every week—it's party central!

3. The flight attendant offers you a pillow and blanket. You
   a. decline—ick, you'd rather not.
   b. decline—you brought your own.
   c. grab 'em and snuggle up.

4. You're heading to the gym after work. Your duffel bag is
   a. hanging in your cubby.
   b. on the floor under your desk.
   c. in the random gym locker where you left it yesterday.

5. You're waiting for the bus. You
   a. stand off to the side—the bus stops up there anyway.
   b. stake out a spot on the bench.
   c. stand shoulder to shoulder with everyone else, jockeying for prime position.

6. The temperature in your office fluctuates like your mom's did when she was going through the change. You keep a sweater
   a. hanging on your coat rack.
   b. on the back of your desk chair.
   c. in your guest chair.

7. You just checked into a swanky hotel. You
   a. stash your suitcase in the bathroom.

    b. put your suitcase on the luggage stand.

    c. drop your suitcase on the bed.

8. The seat next to you on the train is empty. You
    a. breathe a sigh of relief that you won't have to share.
    b. grab the woman you met in line so she can sit with you and talk some more.
    c. think "Score!" and spread out your stuff for easy access.

9. You vacuum
    a. every week, of course.
    b. when your in-laws come over.
    c. um, never.

10. You're headed to the coffee shop on the corner to soak up the free wireless. You
    a. pick a chair by the window and put your laptop bag on the table where you can see it.
    b. sit in a corner booth and toss your bag at your feet.
    c. grab some space on the comfy couch—it's never free!

11. Your six-year-old just walked in from school, dragging his backpack behind him. You
    a. hang it on a hook by the door.
    b. stash it in the front hall closet.
    c. toss it on his bed.

12. A cab, finally! You
   a. hop in and hold your bags in your lap.
   b. toss your stuff in the trunk.
   c. hop in and start going over your day's haul, dropping items on the seat one at a time.

13. You're shopping for work clothes. You
   a. head to the dressing room and hang your stuff on a hook.
   b. head to the dressing room and leave your clothes on the floor while you try on the new stuff.
   c. just try on the new stuff over your clothes— who has time to head to the dressing room?

14. You're flying to Aunt Shirley's house. You
   a. take whatever will fit in your carry-on.
   b. carry on what you can and check your extra bag.
   c. check, like, five bags—you need options!

15. When it comes to vintage stores, there is
   a. no way you'd be caught dead there.
   b. nothing wrong with picking up a few knick-knacks.
   c. no better place to buy furniture to round out your décor.

16. You dispose of your vacuum bag
   a. every time you vacuum.
   b. when it's full.
   c. never—mine is bagless.

## *Scoring the Results*

**Mostly *As*—Bug Hunter:** Maybe you're just super-suspicious—or even a little obsessive-compulsive—but you're already on the right track. You understand that you have tons of control over your life, and your established good habits will make it easier to keep these teeny bloodsuckers at a safe distance. What you learn on these pages will sharpen your instincts and give you hard-core bed bug–fighting chops that you can put to good use to protect your home and your loved ones. (Maybe you'll even be inspired to give vintage stores a try—the *smart* way!)

**Mostly *Bs*—Bugged Out:** Not bad, not bad at all. You don't always make the best decisions, but that's just because you don't know any better. The good news is, you're already rectifying that with the very book you hold in your hands. Soon, you'll know how to prepare for—and clean up after—houseguests, why dumping your stuff on a hotel bed is a no-no, and why that furniture from the vintage store might not be such a great deal after all.

**Mostly *Cs*—Bug Food:** No, I'm not saying you *want* bed bugs to feast on your bod at night (hey, I won't judge if you're into that sort of thing), but it's almost like you've intentionally hung a neon sign around your neck that says The Kitchen Is Open! But don't worry, this book will help you revamp your habits, from teaching you where to put your stuff (hint: never on the floor), to where you probably shouldn't sit (that well-worn coffee shop couch? Not so much), and why the right vacuum, used regularly, can be one of the most useful tools in your arsenal.

## THE TOP SIX MYTHS ABOUT BED BUGS

You wouldn't feel embarrassed to tell your friends if you went for a hike and got bitten up by mosquitoes. You wouldn't feel ashamed to confess that your dog brought in fleas. And you probably couldn't stop complaining if you found out your house had termites. Yet imagine saying to your coworkers, "Hey, I have bed bugs!" (I know, I know—you're mortified just thinking about it.) The myths are so widespread that even though they're not true, they make bed bugs a problem of stigma and shame. Let's start fighting them here! It's not true that...

1. **Bed bugs are only in your bed.** They're not picky about where they hang out—you'll find them in boardrooms, classrooms, the dentist's office, Broadway theaters, and five-star hotels. Oh, and your study, TV room, living room, and...

2. **Bed bugs love a good mess**. Your home or office can be so clean it shines, but bed bugs are just as happy in a spotless place as a dirty one.

3. **Bed bugs are just a poor people's problem.** Tell that to the actor who makes millions per movie but still had to call me.

4. **Bed bugs are the immigrants' fault!** It used to be that bed bugs hitchhiked over the border (and the royal families of Europe, immigrants, and international businessmen all made excellent transports), but now we're passing them on among ourselves.

5. **Bed bugs are only fans of cheap motels.** Believe it if you want, but you're just as likely to be bitten at a five-star hotel as you are at a place that rents rooms by the hour.

6. **Bed bugs are just a city problem**. Some bed bugs love the bright lights of the big city, but their country bumpkin cousins could be waiting for you on your next trip to the 'burbs or the countryside.

## How Do I "Get" Bed Bugs?

Actually, the better question is how do they "get" you and your stuff? You get them because the egg or an actual bug hitchhiked onto your clothes or stuff and made its way into your home—and then laid more eggs.

Bed bugs attach themselves to your clothes, luggage, laptops, cell phones, briefcases, coats, and car seats and car trunks as you go about trying to live a normal life. Even if they don't nab you outdoors, they may get in by crawling between the space under doors or through light sockets if your neighbor, coworker, or hotel next-door neighbor has them. Or maybe they were brought in by a deliveryman, the mailman, the housekeeper, a guest, the nanny, a nurse's aide, your kids, your pet, or even a newly checked-out library book.

Recently I learned about them setting up shop in a very unexpected way. A couple had ordered cabinets for their condo from a national home-improvement chain. When they arrived, I got a call from the building's super asking me to come over ASAP to identify the crawlers that were in the box. I was doubtful that it could be bed bugs, but I was wrong—the cabinets were *covered*. Luckily, I was called in immediately and we were able to eradicate the problem fairly quickly because the super had been educated about bed bugs through his previous experience with us.

## What Do They Look Like?

Behold, the bed bug in all of its various stages of development, magnified about fifty times from its normal size:

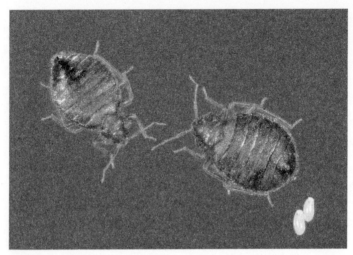

*Left to right: An adult male, adult female, and bed bug eggs. Courtesy of National Pest Management Association/Thomas Myers.*

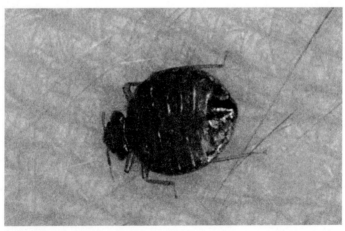

*An adult bed bug crawling on a person. Courtesy of the National Pest Management Association.*

As you know, they are not the only creatures that can take up residence in your house. Any quick Internet search will turn up full-color pictures of other common insects like roaches, fleas, mosquitoes, ticks, chiggers, book lice, spider beetles, and carpet beetles to help you tell the difference between them.

## Places They May Be Living

Use this rule of thumb: if it's large enough to insert something the thickness of a business card, a bed bug can fit inside it, lay eggs, and lie in wait until they can get to you for food. That includes:

- a bed, with headboard, footboard, box spring, and frame included
- in the seams, legs, inside of and under the cushions of a sofa
- a dining room or an office chair
- your bookshelves, books, or magazines (though make sure you are not confusing them with book lice, which are small, colorless bugs that feed on mold and love books)
- light sockets
- picture frames
- laptops
- your handbag
- clothes (even new ones!)
- a child's borrowed toy
- a bathroom vent

- between wooden floorboards
- in wall cracks and crevices
- under and in carpeting
- in the seams of wood furniture, including sofa legs, arms, and bed frames
- under loose edges of wallpaper
- smoke alarm
- behind peeling paint

Yes, bed bugs are hard to look at. Disgusting, in fact. The best thing I can tell you: your chances of seeing a bed bug in your home are slim, as they range from teeny-tiny to the size of an apple seed. How hard? Well, first off, they hide in seams, bed frames, box springs, mattresses, and other hard-to-get-into places, explaining why more than 90 percent of people with smaller infestations never see them. Not even I bother to invest serious time visually inspecting my own home, because I know how hard it is to spot them. Fortunately, there are other ways to detect the presence of bedbugs.

## How Do I Know if I Have Them!?

**The Big(-ish) 3 signs that you have them:** (1) you develop bites, (2) you see bloodstains on your sheets or mattress, or (3) you see their fecal matter on your mattress, box spring, or headboard.

## 1. Bites

Ask anyone what they know about a bed bug and even the most uninformed person will tell you, "They live in your bed and they bite you." Which is actually not true. Yes, they do live in your bed. But they don't bite. They suck, both literally and figuratively.

Beyond that, there are a number of other vital bits of misinformation floating around about bed bug bites. And most of it is on the Internet. Here are the facts you need to know:

**a. Everyone with bed bugs does not get itchy bed bug bites.** To say it another way: you can have bed bugs, but not react to the bites.

Who are these lucky people? The physical appearance of an itchy bite has to do with how allergic you are. So just as one mosquito bite may leave you itching like mad and the other may not itch at all, no two bed bug bites have to be the same. Also, their intensity varies from person to person.

Men are affected less than women by bites. Based on my research, I've determined that nine out of ten men don't manifest allergic reactions to the bites, even though they have been bitten. It sounds unbelievable, but often with couples who share a bed, the woman will be covered and miserable and the man will have nary a bite—even if a professional inspection shows that more of the bugs are hanging out on his side of the mattress. Not only do women tend to be more allergic to bites, but they also happen to be the favored food source. Bed bugs are more attracted to women because of their higher body temperature (warmer bodies represent a blood meal) due to ovulation. Sorry, ladies!

**b. All bed bug bites do not look alike.** If you do have a bite, you may attempt to search online for photos that confirm whether or not it is from a bed bug. Unfortunately, there is no standard-looking bite. Depending where on your body they attack, how long they fed, or how much blood was extracted, there can be anything from tiny red marks, to ball-shaped blisters, to big puffy welts. Welts are typically seen on the torso. Smaller bites will appear on arms and legs.

**c. You cannot tell the moment you get bitten.** When a bed bug bites you, it simultaneously injects an anesthetic, so you feel nothing. For most people, it takes three to twelve hours for a bite to develop, if one develops at all. When a mosquito bites you, you typically know right away. So that itchy feeling you get, the one you may have had the entire time you're reading this, it's not a bed bug—it's your mind.

**d. Bed bug bites appear in close proximity to each other (groups of two or three in small areas of the body).** The classic pattern is a line of three. This is often referred to as the breakfast-lunch-dinner formation. Yet it is not the only way bites will appear. Often people have a bite on their leg, one on their stomach, and another on their neck, and they think it can't be bed bugs because they are scattered. Wrong. If you roll over in the night and dislodge the bug while it's feeding, it doesn't give up and go home, it latches on to a new area.

**e. You may have bed bugs and no bites because you have pets.** What does your dog have to do with it? Bed bugs like dogs, cats, and other mammals just as much as they like you. And your pet is home more than you are. And often not doing much moving around. That makes them prime targets for hungry bed bugs. So while you may never suspect you have a problem because you have not received a single bite, your pooch may be taking the brunt (or all) of them. Don't bother checking your pet (they don't manifest the bites or scratch), and forget using monthly tick applications or giving it a flea bath—these remedies have no effect on bed bugs.

**f. Dermatologists are practically useless.** Common sense would tell you to go to a dermatologist. Yet today's dermatologists didn't spend time in medical school learning about bed bugs or reading about them in their textbooks. That said, a dermatologist *can* help you reduce the itch of a bite by prescribing a topical corticosteroid to ease itching and reduce redness. For a DIY remedy to reduce itching, apply a warm compress directly to the bite. Beware of any dermatologist who wants to biopsy a bite—there is no method of extracting the salivary protein to determine whether it's a bed bug or not.

**g. Bed bugs take a break between meals.** Bed bugs feed every seven to fourteen days. So if you are getting bites every night, that means that a different bug is paying you a visit from the one that bit you the night before. Daily bites mean a very large infestation. Don't wait until you are getting bitten daily to call an exterminator. Early detection is best.

> **Tip:** Use a high-powered flashlight once a week to examine your mattress (both sides), box spring, and headboard for blood stains and fecal matter.

## 2. Bloodstains

In a perfect world, smears or specks of your blood on a mattress would indicate that you rolled over on the creatures and crushed them when they were in your bed. And in some cases, you'd be right. It could also mean that they gorged themselves on you and a little blood spilled out (yes, like crumbs).

## 3. Fecal Matter

Yes, we're talking about poop. Bed bug poop, which looks like pepper spots on your mattress, box spring (especially in and around the cheesecloth covering), or headboard.

Now that you've been forced to consider bed bugs burping up your blood or pooping on your bed, I hate to add that more often than not you can have a bed bug infestation and never see any of these signs. Which is why you need the following additional tools.

### Other Ways to Know if You Have Bed Bugs

**Cups and catchers.** For about twenty dollars you can purchase a device called ClimbUp Interceptors that looks like a cup. You place the legs of your bed or sofa in the middle of the cup. The bugs crawl into it and can't get out. You're left with physical specimens that you can take to a professional pest control operator (PCO) to positively identify and begin the process for bed bug eradication. When this works it's a great tool (cheap and easy to use), but it's far from foolproof and should be one of many tools in your arsenal.

**$CO_2$ detectors.** Bed bugs are attracted to $CO_2$ (which humans and pets emit), and these devices give off $CO_2$ to attract bed bugs. The bugs get trapped, giving you or a professional evidence of an infestation. There are a number of these devices on the market, ranging in cost from $50 to $450. However, they require weekly maintenance in the form of $CO_2$ capsules or canisters, which can cost $10 to $50 a week per monitor. A few caveats: the presence of a house pet can throw off the effectiveness of this machine. And like the ClimbUp Interceptors, it's a useful tool, but far from foolproof.

**K-9 detectives.** By now maybe you've seen cute commercials during the morning news or read about dogs specially trained to detect bed bugs. In fact, certain canines can be a wonderful weapon in determining if you have a problem (with an 85–90 percent accuracy rate)—and in giving you peace of mind. If you have bites, if you have been exposed to bed bugs, or you have reason to believe you may have bed bugs with or without any other telltale signs, get a dog in ASAP. Many homeowners, condos, and co-ops bring in these dogs on an ongoing basis every three to six months, or whatever is affordable. Yes, $250 to $400 per visit isn't cheap, but it's a lot cheaper than full-on bed bug extermination. *And the earlier you catch them, the cheaper and easier it is to get rid of them.* The United States is the leader in successfully using dogs to fight this epidemic. Europe has been slow to utilize them due to quarantine restrictions and slow transport.

### *The 7 Things You Must Know About Bed Bug-Detection Dogs*

1. **These dogs needs to be professionally retrained and tested daily.** This means that there's a risk of the dog not getting the individual attention it needs in an extermination company. Yet having a class A dog enables independent trainers to make money, so they are less likely to skimp on keeping their dogs in top form.
2. **It is best to get a dog that is independent of the exterminator you want to hire to cure your possible problem.** Why? Independent companies have

no vested interest in telling you that your problem is small or big, as they are not making any money on solving it. Extermination companies, however, can make money by exaggerating the findings of their dogs.

3. **You want one who has had the best of the best training.** These dogs didn't just teach themselves how to find bed bugs. So check if the dog and its handler have NESDCA (National Entomology Scent Detection Canine Association) accreditation. Nesdca.com also lists certified dog-handler teams.

4. **If a dog picks up the scent of bed bugs in a certain area in your home,** make sure the dog revisits that space two or three times before the inspection is over and see if it has the same reaction. Rechecking the area will significantly reduce the chances of a "false positive."

5. **If the trainer says it's a packed day for his pooch and suggests another, listen to him.** You don't want the dog coming over when it's too tired to focus on finding bed bugs.

6. **Ask the company to hide a vial containing a bed bug somewhere in your home, so you can see the dog at work.** It also shows that the company has confidence in their dog.

7. **Find out how the dog reacts when it detects bed bugs.** Trust me, you're going to be nervous when he's in your home—why make it worse having a mini panic attack every time the dog barks if its barking has nothing to do with bed bugs at all?

## *What Should I Do Before the Dog Comes?*

- Find someone to watch your pet or put them outside.
- Don't leave food lying around.
- Leave the dog alone, meaning do not try to bond with it, play with it, or distract it from doing its job.
- Do not cook or bake when the dog is there. They need to smell bed bugs, not chocolate chip cookies.
- De-clutter your home so the dog can really get its nose into closets and other spaces. Take down boxes and organize closets.
- Close the windows and turn off fans and air-conditioning at least thirty minutes before the dog's arrival.

If it's determined that you have bed bugs, read chapter 2 followed by chapters 6 and 7.

If you don't, proceed directly to part one, chapters 2 through 5, to learn how to protect yourself ASAP.

# Prevention

In Other Words, Your First Line of
Defense!

# 2

# You Don't Have Them? Great, Don't Get Them!

Staying Bed Bug Free at Home and Work

Most people call me once they realize they've been exposed to bed bugs—through a friend, work, or travel. They may or may not have them, but the experience has left them desperate to find ways to protect themselves in the future. Or they have had them and realized how vulnerable they are to reinfestation. Either way, I'm bombarded daily with questions from people who want to know how to protect themselves—in other words, people looking for ways to prevent an infestation in their homes. As I said in the introduction, beyond telling you how to know if you have them and how to find the right exterminator to get rid of them, my primary goal in writing this book is to help you *not* get them in the first place. *Education* is the key to ending this epidemic, and the more you know about preventing them from crossing the threshold into your home, the easier you'll rest—

and the closer we'll come to getting this menace under control.

I'm going to repeat myself at certain points and I apologize for that in advance, but some of this information can't be stated often enough or clearly enough. And remember, my goal is not to make you paranoid but to raise your consciousness. Once you realize just how prevalent the problem is and just how easily it can become *your* problem, you will almost definitely experience hypervigilance bordering on paranoia. But armed with this book, and the willingness to change a few habits, you can reduce the potential for an infestation in your home by 75 percent. (And if the hypervigilance starts costing you sleep, read about delusional parasitosis—yes, it has a name!—in "It's All Right to Cry" on page 137.)

With nineteen years in this line of work I've seen and heard it all, but I don't walk around in a paranoid state all day. I live my life. I enjoy myself. I take vacations and sleep in hotels and fly on airplanes. I have lots of houseguests and big holiday dinners. And I do all of that with four kids who sit and play and run around everywhere, and then come back into our house. And guess who has never had bed bugs? That's right, me! So every time you think, *Screw it, why am I reading this book!? It's inevitable, I'm gonna get them,* remember that I go into businesses and homes every day that are literally crawling with the bed bugs and I've mastered the steps to keep them out of my personal space and my office. If I can do it, you can do it.

I work in New York City, where practically everyone lives in an apartment building. One of the questions I get every day from people who find out that someone in their building is battling an infestation, is "How can I make my own apartment bed bug proof?" The bad news: it's impossible to 100 percent bedbug-proof an apartment or house. The good news: there are many things you can do to lower your risk of exposure. With roaches, mice, and ants, you can prevent an infestation by sealing your home as well as possible. But bed bugs are a different story. Yes, holes in walls and floors still have to be sealed up, and pipe chases (the holes where pipes go into the wall) still have to be caulked, but then the inside walls need to be treated, creating as much of a barrier as possible with a pesticidal dust. And that's just the beginning. Make no mistake—all of this helps, but even if you caulk the whole apartment it won't keep you from escorting bed bugs right through the front door.

Another danger for apartment dwellers is that you can get bed bugs by doing such seemingly harmless things as sitting in your lobby, leaning on the elevator wall, visiting with your infested neighbor, or just washing your clothes in the shared laundry room. Try not to place your laundry on tables where anyone else puts theirs. Bring your own basket down and move clothes between washer and dryer in that. Once it's all dry, take your laundry back to your apartment and fold it there. People with an infestation often do not realize that they are shaking little critters loose as they unload their laundry bags onto the countertops and tabletops. The same caution should be taken at the gym, at indoor pools, and in playrooms.

And if you *know* there is an infestation in your building, have the K-9 scent detection dog come in and sniff your apartment quarterly, or at least every six months.

One good thing to come out of all the bed bug media coverage is that building managers who used to opt not to do preventive care in apartments contiguous to infested ones are changing their ways. Nowadays, informed managers have adjoining apartments treated as a matter of course. If that's not happening in your building, demand it.

Houses are at somewhat less risk for contaminating each other unless they're joined condos. But you must still be careful of interacting with neighbors who haven't gotten rid of the problem yet.

## At Home

Home. It's where the heart is. It's also where bed bugs are most likely to take up residence (hence the "bed" part) and where you are most vulnerable. Of all the places where you can take steps to ward off bed bugs, this is the most crucial. Happily, it is also the most doable.

The first thing I'm usually asked by people who are afraid of getting bed bugs is "What can I buy?" Shopping is the American way, and there are definitely some useful products out there, which I'll tell you about (along with a few you shouldn't waste your money on). Before I do that, though, I want to stress that the most important thing you can do has very little to do with buying things: it's about establishing a good anti–bed bug routine. But, all right, I know you want to buy something right now, so before I go any further I'll tell you the one thing you absolutely need to have:

A bed bug–certified mattress encasement, box spring encasement, and pillow cover.

(Okay, so it was three things, but they all go on your bed.) Mattress and pillow probably seem obvious to you, but if you have a bed with a box spring, encasing it properly is crucial. Why? Because that is the single place you are most likely to find bed bugs!

Think about it: bed bugs may be small and evil and seemingly invulnerable, but they don't fly, and they need something to hold on to. Box springs are typically made out of wood, usually upholstered on the top and sides, and on the bottom covered with a thin textile called cheesecloth. All of these have tons of grippability—a nice box spring could be home to more bed bugs than you will ever want to imagine. Bed bugs just love them.

But before you buy, remember how I said "bed bug–certified"? For English majors, I don't mean "certified by bed bugs"; I actually mean "certified as effective for use against bed bugs." Here's the thing—lots of places sell bed bug mattress and pillow covers. Most of them offer no more protection than a sheet and pillowcase. Don't just run out and pick one up at the large local chain store—they don't work, and they are overpriced. And please don't be fooled by the giant blow-up photo of a bed bug and all the seemingly authoritative language on the packaging!

The fabric is important, too. It has to be something even the smallest bed bug can't get through, but it needs to breathe so you won't sweat like a prizefighter while you sleep. So forget about rubber, no matter how stretchy, and vinyl will rip. Just because it says "allergen free" or "dust mite free" and has those big pictures of bed bugs doesn't mean it is bed bug certified. And, unfortunately, even "certification" is not always meaningful. Think of the way "all natural" can mean anything a manufacturer wants it to mean; that's about the level of protection you can expect from something labeled "bed bug certified." Most of the encasements we've tested have posed only the slightest temporary challenge to the bed bug of average intelligence.

### So Which One Do I Buy, Jeff?

The only two manufacturers of effective encasements are Mattress Safe and Protect-a-Bed—they are the only cases I have found that will consistently prevent bed bugs from entering through either the fabric or (get ready for it) the zipper. I am so not kidding. Bed bugs in the first stages of life are one millimeter long, and can pass through any conventional zipper, which is what you'll find used on the majority of so-called bed bug covers. You need a zipper with micro teeth and a zipper lock that slides under a little fabric hood. Once you've got good encasements, make sure to check for holes periodically—you don't want to go to all that expense and trouble just to undermine it with a jagged toenail! And PS: even on Protect-a-Bed and Mattress Safe encasements, the zipper must be in locked position to work! It's not unlike installing a home security system and not bothering to set it.

### WHAT'S THE POINT OF ENCASEMENTS IF THEY WON'T STOP BED BUGS FROM ENTERING MY HOME?

There's two points, actually:

1. **Not encasing** your pillow, mattress, and box spring gives your bugs a fertile breeding ground where they can be fruitful and multiply for months before being detected—all while you're spreading them around your home and sharing them with family and friends.

2. **Encasing** makes inspection and detection *much* easier. If you do have bed bugs, you will recognize the bloodstains and other signs against the bright white of the encasements in seconds.

**BEWARE**: A cheap encasement *may* hold bed bugs inside it but be just thin enough for them to bite you through it!

It's got to be an escape-proof microzipper with a proper zipper stop. If the zipper gets backed off even one or two teeth, the parasites can escape.

As I'm writing this, mattress and box spring encasements will run you about $60 each, with pillow encasements about $12 apiece. The Protect-a-Bed products are significantly pricier because, as the first ones to market, they are like the Kleenex or Band-Aids of the industry—you're paying a premium for the brand name.

Now you've encased your mattresses, your box springs, and your pillows. So far so good. But there's a lot more bed left. Do you encase the bed frame? The headboard? The night tables? Your spouse? Well, no—especially not the spouse. But you do have to think about the grippability factor: if you were a bed bug, what would you want to grip?

Metal? Negative. Hard plastic? Negative. Wood? Affirmative. Fabric? Now we're really talking! For bed bugs, wood is a comfy surface: it's easy to walk on, eggs stick to it, it keeps even temperatures, it's in easy access of sleeping bodies...what's not to like? Unfinished wood is the best. Finished wood is a little smoother, a little more challenging, but nothing an enterprising bed bug can't handle. And as far as fabric goes, well, the more intricate the headboard is—pleated, tufted, fancy, carved—the more and deeper the places to curl up and reproduce (see "Special Spots for Inspection" on page 52). If the bugs could write you a little thank-you note, they would. Finally, there are screw holes, which they just love, even though there's metal at the bottom—the rim of the hole is just the coziest little place for Mr. and Mrs. B. Bug to make their little home.

So what does that mean? Do you have to replace all your wooden beds and headboards with something you'd see in a dorm room? You could, I guess, but I wouldn't make it a priority. A lot of bed bug self-protection is about risk assessment, and there is a lot you can do that doesn't involve buying new furniture. One of the biggest ones is cleaning.

Before you tell me how clean you are (or aren't), let me get specific. First of all, bed bugs don't care about that. They're just as happy in an immaculate showplace as in a frat house. (Bright side: you don't have to worry about cleaning before the bed bugs come over.) Second, I don't just mean the kind of cleaning you do when you know guests are coming, or even the kind you pay someone to do. I mean cleaning in a very particular way, with a recommended vacuum cleaner, a recommended dry steam vapor cleaner, and Steri-Fab. Here's how.

**Vacuuming.** Every week you should vacuum your bed thoroughly: mattress (both sides), pillows (likewise), frame, headboard, and all around it. When you're done, whether you have one bed or many, seal, and throw out the vacuum bag immediately—the last thing you want to do is create a family reunion for all the bed bugs in your house. Use a good quality canister vacuum cleaner with a bag (HEPA is nice, but not necessary unless you have respiratory problems or asthma). Two tips:

1. **No-bag is no good!** Bag-free vacuums give bugs the freedom to get into the motor and the body of the vacuum; as if that weren't enough, take out the cup and there they are! I always have the canine inspect the vacuum cleaner; I'm never surprised to find bed bugs inside.
2. **Forget your upright and its crack-and-crevice tools.** Those tools are the secondary function on an upright, which is designed to vacuum floors. The sole function of a canister is to give you equal power whether you're vacuuming the floor, the ceiling, or anything in between. If you're putting all this time into vacuuming, you might as well be doing it with the right tool.

**Dry steam vapor cleaning.** Look for a relatively dry steam cleaner that puts out steam at a high temperature. The estimated death temperature for bed bugs is 130°. The problem is, if you have a steam cleaner that goes to only 130°, when you're vacuuming cracks and crevices you rapidly lose heat as it goes down into the cracks. That's why you want something that goes 180° or higher.

Regular steam cleaners, like the Oreck Steam-It, are great for maintenance, but you don't want too much moisture in cracks and crevices. The good steamers have temperatures so high (upward of 285°) that they evaporate the steam instantly—you can't even soak a towel with it!

You'll find it's actually pretty easy to use, and you should be able to pick up a good one for about $400 to $650. Think of it as a one-time investment, like the encasements, that repays itself with every use. If you use it in flooring, cracks, and crevices when you clean, giving a once-over of the flooring, bedrooms, etc., you will kill bed bugs *and* their eggs, even before you know you have a problem. Doing this regularly will reduce your vulnerability in a big way. Steam vapor cleaning is one of the main weapons shelters, dorms, hotels, and motels bring to do battle against the enemy.

**Steri-Fab.** Steri-Fab is a disinfectant insecticide (pesticide, moldicide, fungicide) spray that I use on an ongoing basis in my home and office. I strongly recommend it. It kills bed bugs on contact (though not their eggs). You can use it around your headboard and on the bed frame itself, on any smooth surface, as well as on mattresses, rugs, carpets, and upholstered furniture. Other than kitchen utensils and waxed surfaces, you can spray it on pretty much anything that doesn't breathe. Follow the instructions on the label.

### Beyond the Bedroom

Good bed bug hygiene in and around the bed is great, but chances are that your home is made up of more rooms than just your bedroom (unless you live in Manhattan). Much of the same stuff applies to the rest of your living space: you still want to vacuum every week, especially the places where you spend the most time (couches, chairs, La-Z-Boys,...). You may want to use the dry steam cleaner in those same places. But there is one place in your home that outranks every other one, the single major battlefront in that war: your front door.

Some days, when the epidemic seems like it multiplied a thousand times from the day before, I fantasize a whole new trend in home design—a bed bug–changing room at the front of the house. Until that day comes, you have to take all the precautions you can to repel the invaders at the door.

Think about it. You've had a full day: a long day at work, followed by dinner in a restaurant and a good movie. In other words, you've been *potentially* exposed to bed bugs all day. You walk in the door. Now do you throw your hands in the air and say, "Oh, well, I guess I just have to take my chances"? Do you run to your bed and roll around on it in your street clothes, daring the bed bugs to do their worst? No. You follow these simple steps and rest easy:

1. You take off your STREET clothes when you come in.
2. You put your clothes in the dryer on high and run it for half an hour.
3. You put your jacket, shoes, and bag in the PackTite.
4. You put on your pajamas and fall into a good night's sleep.

### Wait, Hold On—The "PackTite"? What's That?

I thought you'd never ask. PackTite is the brand name for a thermal heat pack (but it's not a heating pad) or a portable heating unit (but it's not a space heater) that works as a portable de-bed-bugging station. It is a high-temperature heating container, about the size of a steamer trunk when open or a standard suitcase when closed. It can hold anything you wouldn't want to put in a dryer (coats, shoes, handbags, knapsacks, briefcases, etc.). You don't have to empty your bag. Put your stuff in, let the temperature get up to 130°, and keep it there for an hour. If you *did* have bed bugs inside, at any stage of their development, they will not survive the experience!

If I had to say which was more crucial, I would rank the PackTite higher than a steamer. (I wish it were a public company—I would certainly buy shares!) Unlike the steam cleaner, it's pinpoint in its accuracy. Most people haven't heard of it, but it's working its way into homes around America. It's the ideal tool for when you find out your kid has been at the home of someone who had bed bugs, a used book arrives, or you see something you love at the thrift store and bring it home. Just pop it in the PackTite, heat it up, and let yourself exhale.

As good as the PackTite is, it's slow. Think of it like baking a turkey—the more stuff you have in it, the longer it takes. But even better than a PackTite is something you may already have: your clothes dryer.

A clothes dryer is quicker, hotter, and more effective than a PackTite. Its limitation is that you can only put your clothes in it, but it is great for regular clothes! Put your clothes in the dryer and run it for a half hour at its highest heat, ideally 130° (numerous places say that 112°, 113°, or 120° will do it, but research has found that bed bugs were surviving past 120°). By the way, washing machines *may* not kill them, because too often the hot water doesn't cycle through hot enough. If you are looking to conserve space, get rid of your washer first!

Now, do you need to buy *all* of these things? Do you need to buy them all *now*? I would say no. The top buy-now items, as I said, are the bedding encasements—that's nonnegotiable. After that, a good canister vacuum cleaner with a bag, if you don't already have one. Then, if you can afford it, a PackTite. Then, if you're going all out, a steam cleaner. A clothes dryer is great, but if you don't have one (like many apartment dwellers), the PackTite will do what you need. If you have all these things in your arsenal, you may not be invulnerable to bed bugs, but you stand a very good chance of vanquishing them fast if they should ever be so foolish as to invade your home.

Bear in mind also that these are the sorts of precautions you should be taking if you have reason to believe you have been exposed to bed bugs. If you haven't been, or you don't think you have, you may not feel you have to take all the steps above on a daily basis. But if your office is infested, or even if you travel a great deal, often go to theaters or public venues, eat out a lot—things that put you at greater risk for exposure—you may want to consider investing in these things.

You have to know what you consider acceptable risk. I always use the example of the speed limit. It has been absolutely proven that every 5 mph drop in the speed limit saves lives. So a 30 mph speed limit would save thousands of lives a year, but nobody wants to do this—as a society we think that a few thousand lives lost each year are an acceptable risk against the benefits of driving and (hopefully) arriving faster. You always have a choice.

**SPECIAL SPOTS FOR INSPECTION**

Whenever you inspect and clean your home, give extra scrutiny to these known bed bug hangouts:

- wicker furniture
- drapes that touch the floor, bed, or end table
- platform beds and captain's beds
- bed skirts that reach the floor
- the wooden slats under the box spring itself
- popcorn walls or ceilings, or heavily textured wall coverings with cracks or crevices
- the perimeter of carpeted rooms—where the carpet meets the wall

## Bring in the Dogs

Another major thing you can do to make your home less hospitable to bed bugs is bring in the dogs— working dogs that have been intensely trained to do nothing but sniff out bed bugs. At $250 to $600 (apartment to house) a pop depending where you live, it may sound like a luxury, but it could end up being economical. If you brought a dog in four times a year and paid, say, roughly $1,200 over a year of inspections, you would save you a lot if you ever did have them. (And most dog companies will typically give you a deal if you have regular inspections.) Think of it like going to the doctor once a year—you *pray* nothing will turn up during the checkup, but if something did, you hope that at least it won't have gone too far since the last exam. You're playing the odds, striking a balance of practicality and economics. But if you're worried and you can afford it, go to the dogs!

### Don't Bother...

If you go to a hardware store or most bed bug websites, you'll get a lot of information about how you can beat bed bugs. If you follow my recommended practices, you'll be giving yourself every possible advantage. Other than that, let me spare you the bother and expense of a lot of things that *don't* work: So, please,

1. *don't* bother putting things (bag, knapsack, etc.) in the freezer! A lot of people really think putting your bag in the freezer will do the trick, but it won't. If you did have bed bugs in your bag, they would basically go into hibernation and wake up when you brought them out into the warmth.
2. *don't* bother spending hours looking for them— most people never find them unless they're very lucky or there's a bad infestation (if you do, make sure you have a high-powered flashlight and a magnifying glass).
3. *don't* bother with glue boards or sticky traps.
4. *don't* bother having the building guy come in and spray next time he's around. You may be thinking "It can't hurt," but it can.

## How to Handle Letting People in from the Outside

Bed bugs travel in and on people's belongings, so we all have to be extra conscious of the folks we let into our homes. Contractors, nannies, housekeepers, house-guests, and especially overnight guests—what are you supposed to do? It depends. It all comes down to trying to lower your risk without alienating people.

### Invited Guests

When people come to your home at your invitation, should you put them through the same paces you go through on your return from someplace that gives you reason to worry? I would say no, unless you are prepared to lose the friendship, or your guests are as vigilant as you now are. I would personally not ask guests if they had been exposed to bed bugs, though I know people who would. Few people would know or would readily admit it if they did know. Most people view themselves as clean and would be insulted to be asked, though who knows—that may change as the epidemic spreads and awareness grows. In the meantime, if you have any reason to worry, you should wait until they leave and then follow my recommendations on pages 46–48, "Vacuuming" through "Steri-Fab."

Say you're having a few people over for dinner. If your space is small, you *do not* want to have them throw their coats on the bed; if you feel that's your only option, at least put a tarp on your bed, then vacuum and/or treat the area with Steri-Fab afterward. A better alternative is putting hangers on the shower rod in the bathroom. If you have a bigger space, you can buy or bring in a clothing rack and have guests hang their belongings on it.

## Nannies, Housekeepers, Home Health-Care Workers—Oh My!

You have more latitude with people who regularly come to your home to do paid work for you. You can ask (or insist) that they keep a change of clothes at your home. When they come in, they can put their street clothes, shoes, and purses into a sealable plastic bag and change into their work clothes. You may choose to be blunt about it ("There's an epidemic out there, and I don't want you helping these little beasts to infiltrate the sanctity of my home!"), but I always prefer to enlist people's sympathies, even if it means being a little indirect. ("Listen, I'm worried that we may have been exposed to bed bugs, and the last thing I want is to be responsible for sending them home with you, so for your sake..."). And besides, how do you know for sure it's not true?

### Contractors

Good luck with that! People who come into your home to do a specific or nonperiodic job (carpentry, plumbing, electrical work, etc.) are not likely to be willing to jump through the kinds of preemptive bed bug–exterminating hoops that you—or your regular employees—are. ("Thank you so much for coming over at two a.m. to fix my busted water pipe! But wait, before you walk in the door, I just need you to put on this hazmat suit..."). Unfortunately, I have seen too many contractors who we later figured out had escorted these bloodsuckers into the homes of unsuspecting families. The best you can probably hope for is that they are as concerned about getting bed bugs from you as you are from them. That's the best you can *hope* for. What you can do is just be aware of where they go, and treat the area(s) when they're gone.

Remember, contractors and home health-care workers are in other people's homes and/or offices all day, every day. If the problem is in their own home, chances are greater that they'll bring it into yours; if bed bugs are hitching a ride on them from someone else's home or workplace, the chances are somewhat less.

**Profiling**

We've all heard about (or experienced firsthand) how profiling can be misused, but I have to confess that I profile for bed bug probability—*before* I ask people to come to my home. It's not a science (if only!), but I do think about where people would be coming from. Recently, that is; it has nothing to do with their nationalities! Have they just traveled overseas? Do they live in the city versus the suburbs or the country? Are they single with active social lives? If the answer to any or all of these questions is yes, I'm *still* going to have them over (okay, unless they call and say, "I'm on my way from the airport. Hey, did I mention I'm crawling with bed bugs?"). When they leave, if I have any concern, I'll do the regular vacuuming/steam treatment/Steri-Fabbing. And then I'll let it go.

I know a woman, a professional in her late thirties, whose bed bug infestation I treated (I was her third and, thankfully, her last pest control operator). It's an understatement to say bed bugs had traumatized her for eleven months. Now that she's bed bug free she refuses to let friends stay with her if they're coming from a hotel. She tells would-be guests why she feels this way, and after the bed bug summer of 2010, friends have definitely been more understanding of her boundaries. I personally think she's a bit extreme, but I respect that she's open, forthright, and aware of how much risk she can tolerate. If you're worried about bed bugs, you need to figure out how much risk you can tolerate and act accordingly. This woman would rather irritate a friend than risk losing a friendship entirely if the friend brought bed bugs into her home. Who can blame her?

## If You Have Pets

Pets and bed bugs: not a great combination (except for the bed bugs). Humans are generally the host of choice, but if bed bugs find other warm-blooded creatures to snack on, they're not fussy. They'll go for cats, dogs, birds, hamsters, gerbils, ferrets, you name it, and they'll dig in deep. If you don't see bites on your pet, it doesn't mean anything—bed bugs can be biting pets for a long time before they start biting you. Bed bugs are not like ticks or fleas, so don't bother applying Front-Line or giving your pet a flea bath; it won't have any effect.

If you suspect, you need to inspect furniture (especially pet beds) and toys for the telltale signs of blood spots, fecal stains (which look like pepper stains), and bed bug carcasses. Some people say they can recognize the smell, but speaking as a professional who can tell the presence of a roach, mouse, or rat infestation just walking in the door, my nose can't detect the presence of bed bugs, and I have doubts about anyone who thinks their nose is better—unless they are dogs. Get the animals out of the house and see if there's biting; if there is, bring in the dogs (the bed bug–sniffing kind) to inspect.

Why would bed bugs bite dogs and not you? Unlike you, your dog is typically around nonstop and sleeps about eighteen hours a day—that's a good long time for any bed bug to take aim and hit the target. But bed bugs need to be near a feeding source of any kind—human, mammal, or bird—so that doesn't mean only you and household pets. If you keep farm animals, that works fine for them. For example, they've become rampant on organic chicken farms—I have seen them there in piles, mounds, of biblical proportions. That doesn't mean you shouldn't raise organic chickens if you want to. Just be aware!

## Why Are You *More* at Risk if You're Single?

There's nothing wrong with being single, despite what you may have heard from my mother. Not being part-nered probably means you have an active social life, which is great (sorry, Mom)—but it does put you at greater risk for including bed bugs in your social cir-cle. Statistically you increase your odds just by going out more than us stodgy old married types. Multi-ply those odds if you have a roommate, and multiply again if you have more than one. Keep multiplying for each intimate partner (remember the *bed* part of *bed bug*?). You could look at bed bug transmission as the latest advance on STDs: not only do you have to worry about who you're sleeping with, and who *they* slept with, and so on—you don't have to be intimate to get the "benefits" of transmission.

I once treated the apartment of a man who kept getting bed bugs even after his apartment had been cleared by a canine inspection. I grilled him and his girlfriend again and again about where they hung out, if there was anyone in their lives who had bed bugs—no, nobody, it was a total mystery....It turns out that the guy had not one but *two* other girlfriends besides the one he was living with. One of these women (and it only takes one) had a roommate with bed bugs. Mystery solved! (The relationship? Not so good.) The point is, the more contact you have with people's space and belongings, the greater your chance of getting them. Don't be alarmed, just be more aware.

PS: Five years later, the same guy called me—he had bed bugs again. My first question: "So how many girlfriends do you have these days?"

### *What You Can Do About It (Besides Stop Playing Around and Find Someone, Already)*

I'm not trying to scare you into settling down ("Let's get married so we can cut down our chances of getting bed bugs"), but abstinence will help your body and certainly your soul! You can keep the odds in your favor by applying the same basic measures I recommended at the beginning of the chapter (mattress and pillow encasements, vacuuming, steam cleaning, Steri-Fab, clothes dryer, PackTite). You just have to apply them more often!

## When Your Home Has Other Little Critters (Namely, Kids)

People often ask me if there's a routine that will minimize their kids' chances of exposure. How can my child avoid bed bugs at school, on play dates, at summer camp? How can I teach my kids to be smart without making them pint-sized paranoid crazies? Should I throw out all the toys—bed bugs love toys, don't they?

Well, your children probably don't need as many toys as you (or they) think, but bed bugs really like people. They are exploding at schools, and they are being seen more and more on the summer camp scene, but let's pause the panic button: bed bugs are epidemic, but they are not the bubonic plague! More and more camps bring dogs in as early as May, but if your child comes back from sleepaway camp, put all his or her clothes in the dryer on high for a half hour. Non-dryer-friendly belongings go into a PackTite, if you have one; otherwise, spray Steri-Fab and vacuum. If you have particular reasons for concern about your child's school or a playmate's home, you can institute a ritual strip-down-at-the-door, clothes-in-the-dryer, shoes-and-book-bag-in-the-PackTite procedure, but if you're doing it "just to be safe," you're sowing the seeds for some serious bug phobia (not to mention adolescent rebellion).

Same thing goes for toys. In the past, I heard count-less mothers whose homes were infested cry about how they had to throw out their children's favorite toys, or put them in three layers of Ziploc bags (the toys, not the kids) and then hide them for up to eighteen months. They could have just tossed plush toys in the dryer for an hour. Now, with the PackTite, you can debug pretty much anything in your child's collection, other than Play-Doh.

## When Your Kids Go to College: What to Do Before Your Kid Moves into the Dorm

According to *USA Today*,* college dorms throughout the country are plagued by bed bug problems, due to the social nature of dormitory life. From what I've seen myself and heard from others in the business, the college dorm that is *not* battling bed bugs is approaching nostalgia status ("Back in my day, we never had to worry about 'em!"). The close proximity of rooms, and students' tendency to share clothes, items, books, and each other, has made it the perfect breeding ground.

So what do you do before your kid sets one foot back in your home? Basically, you treat him like a new prisoner getting deloused at the door (no body cavity search necessary): clothes into the dryer, suitcase and contents into the PackTite. The real issue with college kids is their belongings, desks, any kind of furniture that they're keeping from semester to semester—used textbooks, etc. I suggest asking them to store furniture and books at school and bring only clothing home.

## When You're Moving

For starters, you can always ask if the house or apartment you're moving into has had bed bugs in the past. In New York State, landlords are required by law to provide a rider to a lease indicating which apartments in the building have had infestation in the past twelve months (they can be sued if they conceal the information and you end up infested), and other municipalities are looking at doing the same. With or without full bed bug-disclosure laws, you should definitely bring in the dog(s)! While dogs are not foolproof (they can't sniff out things that are substantially above their heads—like an infested stucco ceiling), they are still the fastest and relatively cheapest way of finding out what you're moving into.

So now you've found your new home, the dogs gave it the all-clear, and you can devote your energies to packing up and moving without fear of bed bugs, right? Wrong! Moving trucks are one of the easiest ways bed bugs travel! You may move from your bed bug–free home to another bed bug–free home and find you have bed bugs. To quote Charlie Brown, "Arrgghh!" Many people don't get rid of their bed bugs and instead try to run away from them. As a result, their bed bug–infested belongings end up infesting the trucks they move in and passing along the problem. Moving companies claim to fumigate their trucks, but what they call fumigation is at most setting off a bomb (a complete waste of time). If companies are doing anything, they are certainly not doing it between every move, and when they do it at all, they are not likely to be effective. They are certainly not laundering the tarps every single night, if ever.

My tip: If I were moving, I would take a thirty-hour detour, and fumigate the truck with all my belongings inside. A qualified company will tent or seal off the truck and pump it full of Vikane gas (sulfuryl fluoride), killing bed bugs in all stages. It is added time (pickup, twenty-four hours to fume and aerate, then redelivery) and expense (about $1,500 per truck), but when you think about what it costs to move, and what home means to you, it's probably worth the expense to have that peace of mind.

## You Have to Leave the House Sometime!

By now you're much more aware of the bed bug problem than you ever were (and definitely more than you ever wanted to be). Let's talk about some simple steps you can take to protect yourself while you're out and about: don't throw your coat or jacket in a pile of coats at a party (see page 54); don't put your bag on the floor when you go out to dinner; don't sleep at the home of a friend who has bed bugs.

Bed bugs are not like ticks—the chances of one hitching a ride on your person are slim. Your handbag, coat, backpack, laptop bag, sweater, and suitcase provide a much cushier, safer ride. It's not like they're standing on the corner with their thumbs out and hopping in when you stop at a light. They just seem to sense that your bag is more comfortable than the surrounding environment, and being the rude buggers they are, they come right over, uninvited. Sometimes they mosey inside and don't crawl out before you pick it up, other times they just strap on their seat belts and hope to survive the ride on the outside of your bag.

Lucky for you, they're like a four-year-old who crawls under his bed every time he plays hide–and–seek: predictable. Here's where they are most likely to be; check these spots before you carry bags into your home. They're very difficult to spot, but this is where they are.

**Seams.** These are easy to grip and make for excellent hiding places.

**Pockets.** They crawl in, they don't crawl out.

**Zippers.** The space between the fabric and the zipper? Genius place to take up residence.

**Fabric.** Seriously. They can just hang on to your bag for dear life.

---

### THE MATERIALS BED BUGS LOVE (AND HATE)

Want to make a bed bug looking to latch on to something very happy (well, as happy as a bug can be)? These are the materials that allow for a lot of grip:

- cotton
- wool
- wood

And these are the ones that give the buggers quite a bit of trouble:

- nylon
- plastic
- iron
- glass
- polished metal, stone

---

I can easily tell you to be careful of all public spots: movies, restaurants, theaters, etc. But you can make yourself as crazy as Jack Nicholson in *The Shining* (or *One Flew over the Cuckoo's Nest, Batman, As Good as It Gets*...pretty much any of his movies). When you go to the movies, do you put down a plastic bag? When you go to the doctor, do you sit only on the leather chairs and not the fabric ones? When you go to the restaurant, do you spray Steri-Fab on the seat, stand around for ten minutes while it does its work, and watch other patrons lose their appetites at the smell? No, you don't—you'll drive yourself crazy and convince everyone else that you are, too. Stick to the stuff you can control, like your home and where you work. I'm giving you this info to make you feel safe, not neurotic.

And this is really the number one rule: you can't control what's out there, but you can absolutely control what comes into your home. Take some precautions while you're out, but don't make yourself nuts. And whenever you think you may have been exposed—at the movies, on a trip—make your most vigilant moments the ones you spend at the threshold of your home. My wife and I travel, we go to the movies (okay, not really, but it's not because of bed bugs: I'll just say "four kids" and let it go at that), we go to the theater, we let the kids have sleepovers—all because the only real point of concern for us is our own front door.

## The Bottom Line

You're lucky: you don't have bed bugs. And now you know that you don't have to depend on luck *not* to get them, there is a good bunch of preemptive measures you can take to keep the microbarbarians outside the gate. To review:

- bed bug–certified mattress, pillow, and box spring encasements
- regular vacuuming
- steam cleaning
- Steri-Fab
- clothes dryer
- PackTite
- bed bug–sniffing dog inspections
- front-door protocol

And, if you're really feeling your oats, you can make a little sign to hang on your door:

The Bug Stops Here.

# 3

# Planes, Trains, and Automobiles

How Bed Bugs Hitch Rides from
Other Travelers to You

If you've gotten this far in the book, you're likely more
than a little freaked out; locking yourself in your bed
bug–proof fortress is probably sounding pretty darn
good right now. But that's no way to live. Whether it's
a crowded bus ride to the mall, a rejuvenating winter
jaunt to an exotic island locale, or an obligatory train
ride to see your in-laws (probably not so rejuvenating),
you've gotta get out at some point. The problem is,
while you may have paid to use that seat and think of
modes of transportation as vehicles to get humans from
point A to point B, bed bugs are skilled stowaways who
use luggage holds and subway seats just as often as you
do. And they are just itching to latch on to your stuff
and join you on an adventure that ends in your home.
But if you follow these suggestions, you'll be much less
likely to pick up any unwelcome souvenirs. And that,
ladies and gentlemen, is where I come in.

It is indeed possible to travel in this mad, mad, mad,

mad bed bug world. Exhibit, let's call it "D" ("A" is so overdone): A friend's wife happily boarded a plane to India. She was going to an art show opening, but when she deplaned fourteen hours later, all she could think about were bed bugs. Not because she was paranoid, not because she had an inexplicable love for creepy crawlies, but because she was positively *covered* in itchy bites from the bugs that had taken over her plane seat. Lucky for her, she'd fought them and won when they tried to take over her spacious Westchester home four years before, so she recognized the bites and knew exactly what to do to keep bed bugs from ruining her trip—or hitching a ride back to her house to stay. Stick with me, and you'll know, too. In this chapter, I'll tell you everything you need to know to travel safely.

## The Short of It

While it's the long trips that probably freak you out most, it's the short, daily jaunts that provide the most frequent chances to come into contact with bed bugs while traveling. Here's how to keep yourself safe:

- **In a taxi:** Keep your bag on your lap if possible. Never take things out of it and put them on the seat. Any bugs that were left behind are just waiting to hitchhike their way back to somewhere that will provide them with their fill of blood. If you have to take off your hat and scarf, keep them on your lap— and off the floor and the seat. If you have luggage (and are extremely cautious and don't give a lick that people think you're crazy) you can put it into a garbage bag and tie it tight to protect it from anything someone else dropped off along the way. Or you could skip the bag lady look and just be extra careful when you bring your luggage back into your home (see page 80 for how to do it).

- **On public transportation:** You can pick up more than a nasty cold from the guy with the hacking cough; after years of telling my clients to be extra vigilant about bed bugs when using mass transportation, a New York City official indeed confirmed that bed bugs had been spotted on board trains and in stations. I don't advise sitting on the benches while you wait; I've definitely found bed bugs on them. The best place to be? Standing, holding on to a pole, not touching anyone else. If someone boards wearing a huge coat or carrying a bag that keeps brushing against your sleeve, move to another pole (believe me, I know this can be difficult during rush hour!). And if you see some dude pulling everything out of his bag and dropping it on the seats—go to the other end of the car. You never know what he could be leaving behind for the next person who sits down.

## Your Home on Wheels

If you live any place other than a major city, your car is probably your most frequent mode of transportation. While our flat-bodied friends may turn up their noses at taking up residence in your vehicle (they heard the food service is terrible), they're not above passing through. Think about it; if they are in (or on) your shopping bag or purse, and you place it on the passenger seat next to you, there's a good chance you could leave one behind when you get out.

The bad news is that as hard as they are to find in a bedroom, they are even harder to find in a car, so there's no real effective inspection you can do—there are lots of nooks and crannies, and likely no fecal matter or bloodstains to spot. The good news is, there are simple, unobtrusive things you can do to protect your home on wheels. (If you suspect there are already bugs hiding out, have a dog sniff out the situation now.)

## CHECKLIST: Keep Your Car Pristine

They key to preventing your car from becoming a rolling bed bug haven is to treat it like an extension of your home. Here's what I recommend:

- Vacuum it well every couple of weeks, just like your home. Be sure to get the seats, floor, mats, and trunk. Hit the seams well, and if you have cloth seats you'll need to take more care than if you had leather ones. Dispose of the vacuum bag when you're done.
- Spray Steri-Fab in all of the cracks and crevices that you vacuumed. Then let the car aerate before taking a ride—it will have a smell for approximately fifteen minutes.

- Line your trunk with a tarp. It'll give you an extra level of protection when you throw in suitcases, say when you pick up a friend from the airport. Just pull it out and spray it down with Steri-Fab before you vacuum.

- This is optional, but will give you an extra measure of protection (and perhaps peace of mind). Keep a box of dollar-store garbage bags in your trunk. Then slide luggage into them, and tie off the bag. Anything that's in or on the suitcase will be locked inside—and out of your car.

- In the summer, you can take advantage of the weather: Leave the car out in the sun on days higher than 90°. With the windows up, the car will reach a core temp of over 120°, which is hot enough to kill them or drive them out after an hour or two. If it turns out that you do have an infestation in your car and it's not summer, you should have it fumigated.

## Planes, Trains, and...Buses

Now for those pesky long trips. They might take you somewhere fabulous, but getting there can wreak havoc on the bed bug–worried mind: hours sitting still, shoulder to shoulder with strangers, your bags crammed in next to theirs in some dark, dank cargo hold. I hate to say it, but you should basically assume that you're going to come into contact with bed bugs when you travel. Take that slightly pessimistic attitude, and you'll do everything in your power to avoid bringing them back to your home, because as I've said before, your first line of defense is a good offense at your front door. And while traveling, here's how to be careful.

## CHECKLIST: Your Travel-Safe Tool Kit

The first step to a bed bug–free return is to make a few key purchases:

1. Hard-sided luggage—it's harder to latch on to than cloth and provides fewer places to hide. I wouldn't recommend buying new luggage for the sake of these critters, but if you're in the market for a new set, this is the way to go.
2. Rest Easy. It's a bed bug repellent that uses permethrin to keep them from crawling onto (and into) your bags.
3. Steri-Fab. You're definitely going to want to use this at home. If you're going to carry it on a plane to use at your destination, pour some into a TSA-approved spray bottle to avoid having it tossed in the security line.
4. LED flashlight for room inspections.
5. Disposable gloves in case you come in contact with blood spots or fecal matter during your inspection.
6. Garbage bags to put your baggage in, and Ziploc bags for personal items.
7. More garbage bags for inside your luggage, or bed bug–proof luggage encasements. I like BugZip

brand; right now, a large one will run you about twenty dollars. (I love disposable laundry bags for handling infested items in the home, but I wouldn't suggest using them for travel. There is just too much room for error—they can't get wet, they can't go directly into the dryer—and the goal is to make this as easy as possible.)

8. A digital camera or camera phone that takes clear shots; if you find any bed bugs on your trip, use it to snap photos just in case you need to prove the source of an infestation to recoup your expenses from the hotel.

### How to Pack

Now here's what to do with them:

1. Spray your luggage inside and out with Rest Easy to deter bugs from crawling into or onto it. It's toxic for birds and cats, so take it outside if you have them in your household. Let it dry for ten minutes before you touch it or put any items in it.

2. Place a garbage bag or luggage liner inside your bag, then add your clothing and other stuff. Tie off the bag when you're done packing. (You can do the same for your laptop so that it's protected inside its bag; many models can be heated in a PackTite when you return home, but if the manufacturer advises against it—or you're working with a dryer—using a plastic bag now can save you headaches later.)

3. If you want to put items in the outside pockets of your luggage, first place them in Ziploc bags.

4. Use additional garbage bags to encase and protect your luggage (this is a step favored by the most hard-core and could serve you well when storing bags under buses and in taxicab trunks).

5. When you get to your location, use your flashlight to do the inspection detailed in the box "Step-by-Step: Hotel Room and Cruise Ship Cabin Inspection" in chapter 4.

6. Then spray Steri-Fab in areas you've inspected, if you decided to carry it with you (this is optional, but might give you some additional comfort).

7. Be sure to retie your garbage bags (or completely zip your luggage liners) each time you close your suitcase.

8. When it's time to go home, use additional garbage bags to protect your things on the return journey.

9. Follow my protocol for returning home on page 80.

### CHART: Take It or Leave It?

Take:

- necessities only.
- clothes that can withstand high heat.
- your flashlight.
- garbage bags, BugZip suitcase encasements or drawer liners, or garment bag encasements.

- your small bottle of Steri-Fab.
- a digital camera or a phone with a clear camera.

Leave:

- things that need to be dry cleaned (unless you're into spending money).
- lacy lingerie (heat could destroy them).
- delicate vintage clothing (ditto).
- extraneous electronics, such as travel alarm clocks (they are hard to treat).
- anything that you'd want to curl up and die if it was destroyed.

### To Carry On or Check?

Big question, huh? When it comes to airplanes, checking your bags can cost more than you think. Besides the fact that many air, bus, and train lines now charge additional fees to check bags, handing them over is like drawing bed bugs a map to your front door. Why?

Because the less control you have over where it goes, the less you can protect it. Check it, and there are countless chances for it to be exposed to bugs: leaning against other people's bags while waiting to be scanned, riding on those rickety transport carts with the fraying carpet, knocking around during the ride, bouncing all over that conveyor belt at the baggage claim at O'Hare or Atlanta International Airport. And all the while, bed bugs are potentially crawling from one bag to another, exploring and claiming your stuff as their own, and dropping eggs like little improvised explosive devices set to detonate in your bedroom. (If only airport security could spot bed bugs with the same efficiency that they find your tucked-away bottle of Snapple!)

If you're taking the bus or train you probably have some leeway, as regulations aren't nearly as tight there. You could put your luggage in garbage bags before you stow it beneath, and you'll have superior protection—and free hands. Just trash the used garbage bag (in a can outdoors) when you grab your bag at your destination.

### *Now What Do I Do with My Stuff?*

Nothing is foolproof, but I pack as light as I can, and use a hard-sided suitcase or a simple waterproof nylon knapsack (harder to crawl onto and into) for short trips. I try to fit it under the seat in front of me to reduce the chance of it touching others' bags, but that's only a small protection over putting it in the overhead. If I were the sort to carry a handbag, I'd keep it in my lap.

## What to Do if Your Chosen Mode of Travel Is Infested

Picture it: you're on a plane, and you see what you think is a bed bug scuttling in the crack between your seat and the side of the plane. (True story. I have clients who have seen this and called me from thirty-five thousand feet.) What do you do? If you have a camera within reach, snap a photo—seems wonky, but it will help you file a complaint or take action if you find that you've carried its many-legged cousins into your home. Then push that button overhead, tell the flight attendant what you saw, and ask if there are any open seats on the plane you can move to. (Obviously, if you're on a mode of transport without assigned seats, you can move yourself.) Let someone at the airline know when you disembark, either at the airport or by phone. But I'll be honest—these things are hard to follow up on, because there's no way for you to know if and how they actually took care of it. Remember, their priority is to keep planes in the air.

## WHAT DO THE INDUSTRY FOLKS HAVE TO SAY ABOUT ALL THIS?

In a phrase: not much. I spoke to folks at the industry associations that represent the major modes of travel in the United States, and not one of them told me that they have a bed bug–prevention program in place. Not one. Some representatives swore they would establish a comprehensive set of best practices if one of their member organizations ever reported an infestation, but *we* know that when it comes to bed bugs, the best plan is a proactive prevention plan—not a reactive one after an infestation has been identified.

What can you do about it? Well, for one, be vigilant, using the tools in this book. For two, apply pressure to these folks. If we were all calling airlines and asking if they regularly inspect planes for bed bugs, they may start to pay attention or risk losing customers. In fact, a good bed bug–prevention protocol is a wise marketing tool at this point; hold them accountable. The phrase "vote with your dollars" is a cliché for a reason: it works!

## Road Trip!?

Figure it's safer to rent a car and hit the open road? You're probably right. But you still need to take some precautions; heck, at least a dozen people probably drove it over the past month, and who knows what they brought along for the ride? You could ask the rental agent if they've found any signs of bed bugs, but they probably won't know, or won't want to tell you. No, the name of the game here is to have a healthy degree of skepticism and assume there's something there, whether you see evidence or not. So pack garbage bags or a tarp, and use them to protect your suitcases and other bags from anything that the previous driver may have left behind. And when you get home, do all the stuff I recommend in the following checklist to keep bugs from crossing your threshold.

## CHECKLIST: Before You Reenter Your Home

As I keep writing, the real trick to staying bed bug free is doing everything in your power to keep them from making it inside your home in the first place. Run through these steps, and you'll reduce your chances of bringing those little parasites into your home by 75 percent. It may seem inconvenient at first, but this method has been tested by hundreds of previous clients who were repeat victims, and the ones who follow it have had no recurrences of infestation.

- Strip down at the door. Put all the clothing you're wearing and all the clothes in your suitcase in a plastic bag. The bag is a must—you don't want to drop bugs on your way to the laundry room.
- Put everything else in the PackTite, if you have one. Otherwise, leave it by the door.
- Empty everything in the plastic bag into the dryer. Set it on "high" for thirty minutes. Note: Small 110-voltage apartment dryers may not get hot enough to kill bed bugs.
- Toss the garbage bag in the trash—outside.
- Vacuum, then spray your luggage (and stuff that couldn't be treated with heat) inside and out with Steri-Fab.
- Vacuum the area where you stripped and remove the bag.
- Dispose of the vacuum bag outside of your home.
- Spray the area with Steri-Fab.

## The Waiting Game

In general, unless you see a bug come crawling out of your suitcase doing the Macarena, it will be two days to six weeks before you know if you carried bed bugs home with you. You will know because you'll have those itchy bites (unless you're the type who doesn't manifest them), or see blood or black fecal stains on your sheets. As I've said before, if you have pets, things are a little trickier. Bed bugs could set up residence in their sleeping areas for months before they make their way to yours, so check their spaces regularly (see pages 57–58).

## Legal Matters

So of course the big question everyone wants answered is: "Can I sue if I pick up bed bugs while traveling?" I turned to Eric Baum at Simon, Eisenberg & Baum (and, no, we're not related) for a straight answer. Basically, it all comes down to one term, and whether or not you can prove it. That term is *negligence*.

Here's how it works: Say you think you carried bed bugs home after a cross-country flight. You can sue if you can establish that the airline was negligent. This means that the airline had a duty to provide bug-free travel to its customers, and they either knew (because they had been notified) or should have known (through the normal course of business) that there was a bed bug infestation, and then failed to act in a reasonable and timely manner to rectify the problem, and you suffered damages as the result of their inaction.

But practically, it is difficult to prove negligence. For one, it's hard to establish that the bugs came from the plane, as opposed to your hotel, the restaurant you ate at before your flight, or the cab ride you took home from the airport. But let's say you actually saw a bug crawling on your seat and took a photo of it on the plane. That photo proves that there were bugs on the plane, but now you have to prove that the airline knew they were there and didn't take reasonable and timely steps to eradicate them. To do *that*, you need to show that others had filed previous complaints about that aircraft and yet the airline did nothing.

That said, *if* a number of people complained and the airline knew about it and did nothing and you could prove those things, you could bring a successful claim and win funds to cover extermination bills, doctor's charges, pain and suffering, and even punitive damages if the conduct was egregious.

## The Bottom Line

The world can seem like a scary place once you know there are bed bugs in it. But that's no reason to become a hermit. There are simple things you can do to keep those tiny vampires at bay. To review:

- Assume you're going to come into contact with bed

bugs on your travels, then act accordingly (pages 73–75).

- If Confucius had been plagued by bed bugs, he would have said: He who controls his stuff, controls his exposure (pages 76–77).
- Check—and clear—bed bugs' hiding places before taking bags into your home (page 80).
- Treat your car as an extension of your home (pages 71–72).

# 4

# The Traveling Life
# (Yours and Theirs!)

Staying Bug Free in Hotels, on Cruise Ships,
and at Other Homes Away from Home

Fun stories, great pictures, new friends and recipes, and
a little peace of mind—that's what you're supposed to
bring back when you travel. *Not* bed bugs. Yet sleeping
away from home (especially at hotels) is a common way
to catch bed bugs: unsuspecting hotel guests bring them
in, and many properties (no matter how many stars
they boast) do not have suitable prevention and treatment programs. And most homeowners aren't aware of
infestations until it's too late, so your buddy can't give
you a heads-up before you show up with your suitcase.

But there are things you can do, even if you find
yourself in a place where bed bugs have already moved
in. I've been there. I went to a pest control conference
at a huge hotel in Atlantic City back in 2010, and as
I was checking in I saw several of my PCO buddies
walking from the elevators to the front desk, luggage in

hand. Why? They'd found bed bugs in their rooms! It turned out the hotel had been treating an infestation on four different floors for more than three weeks but had continued to sell the rooms. Awful, right? But because these guys knew exactly what to do (and what not to do) when they got upstairs, they were able to save themselves a ton of grief. (How ironic would it have been if a bunch of pest control company owners got bed bugs at a pest control conference?) In this chapter, I'll let you in on what they did—and how you can use those tactics to keep your sanity intact.

## The Thing About Hotel Bed Bug Policy Is...

...it doesn't exist! No, really. While state health departments create and enforce sanitation standards for hotels, and some hotel franchise agreements require operators to maintain a pest-free environment, there is no industry-wide policy or even recommended best practices for dealing with bed bugs, according to the American Hotel and Lodging Association. Why? They see it as their role to educate hoteliers, not regulate them. And the result is that there is no standardized procedure for preventing and treating infestations. On the individual hotel level, pest control companies are called in to handle problems once they are discovered, but there is no established protocol, and many companies aren't up on the latest methods. Add that to the fact that some hoteliers opt to not treat reported rooms because they are loath to lose money when they take them (and goodness forbid, the adjacent rooms) out of service—or they place the rooms back in service too soon without having them cleared by bed bug–sniffing canines against the counsel of their pest control professional—and you have a perfect bed bug storm. Unfortunately many hotels have to be busted on the news or sued for negligence before

## HOTELS: THE PERFECT (BED BUG) STORM

I've treated more hotel rooms than I can count for bed bugs, and I have to say all things conspire to make them havens for bed bugs. There's a transient guest population that could be bringing in a new infestation at any given time, and there are usually hundreds of employees who could be transporting them in both directions. There is a culture that often emphasizes speed over safety, so infested rooms are often sold before they get an all-clear from a bed bug–sniffing dog, and the housekeeping staff is often encouraged to turn over rooms quickly, leaving little time to inspect. And about that housekeeping process: sheets are thrown on the floor, and clean linen carts can sit side-by-side with dirty ones, which provides infinite possibilities for freshly laundered sheets to carry bugs from one room to another. It's no wonder many hotels are crawling with bed bugs, and they are incredibly difficult to treat effectively.

they call in someone like me to create their prevention and treatment protocol, which needs to incorporate housekeeping, the engineering staff, bellhops, and those working in the laundry room.

## So What *Should* Hotels Do?

We could all sleep a little easier if the industry as a whole, and individual hotels especially, worked with their pest control companies to develop prevention, identification, and treatment plans that take into account the latest technology. In general, here's how that policy should look:

- Properties should have someone on staff, probably someone in the maintenance department, who inspects rooms at regular intervals. While hotels are easier to inspect than private homes because they are generally streamlined and offer fewer hiding places, it's not realistic to have housekeepers do it every time they turn over a room when guests are waiting.
- If bed bugs are suspected or found, the room should be taken out of service immediately, and the call should be made to the pest control company.
- Take apart the bed and other furniture in the room to inspect for bugs.
- Inspect adjacent rooms. If they are cleared, they can go back into service.
- Treat with a full-service program.
- Leave the room out of service for fourteen days to allow for follow-up treatments.

- Have dogs inspect after treatment to clear the room. Only then should the room be put back into service.

## Do Your Own Recon

You can't hide from hotels forever. Someone you love is going to get married out of town, your job is going to send you to meet with clients in some far-flung corner of the earth, or your wife will want to take that vacation you promised her, and you'll be forced to leave your cocoon. You'll be okay, I promise. You just have to add a step or two to your usual hotel search process; 3.5 stars on Hotels.com isn't going to cut it anymore. Many hotels will not exactly be forthcoming if they've had a problem, so after you identify a place that *looks* great, it's time to run a search for its bed bug status. There's no formal, government- or industry-mandated reporting system for hotels, but I suggest you check out each hotel you're considering on these sites *before* you make a reservation (just remember that they are a great starting point, but far from extensive):

- **BedBugRegistry.com:** This site was created in 2006, and is a free, user-driven database of places where bed bugs have been spotted in the United States and Canada. There's no way to be entirely sure if bugs are still alive and kicking, but if you have the choice of staying at a site someone listed and one someone didn't, I think you know which one I'd recommend. You can search for a particular hotel by name, or just put in a city and state and check to see what properties have been listed. Listings include the date reported and, usually, the personal stories of the people who reported it. Posts are removed after two years if no one else reports a sighting.

- **TripAdvisor.com:** While this site was started in 2000 to let travelers advise other travelers on how to make the most of their trips, users started adding info on bed bugs to their reviews several years ago, making it an invaluable resource for researching before you type in your credit card number. Do a search for "bed bugs" and the city you want to stay in. (You can add the name of the hotel if you already have a place in mind.) Read the reviews carefully, taking special notice of the dates travelers posted. (Your alarm should also sound if you see reviews complaining of blood spots or black flecks on sheets.) Includes hotels all over the world.

### The #1 Question to Ask Before Booking a Room

Finished your research and found a well-reviewed spot at a decent price? Whether you're staying in a hotel, or cruising the Caribbean, there's just one more step to take before you click Book It. No one will tell you if they have a current infestation, but this trick works. Call housekeeping directly, *not* the front desk, and say: "I know there's probably nothing now, but have you treated for bed bugs in the past six weeks?" If they say yes, find another place to lay your head.

## DOES IT MATTER IF I STAY AT A FIVE-STAR HOTEL?

...And six more questions with the answer "Not so much."

1. **Does it matter if I stay at a five-star hotel?** Not so much. I'll say it again: bed bugs aren't classist in the least. You're just as likely to pick them up while lounging on organic Egyptian cotton sheets as on threadbare ones.

2. **Will bringing my own towels help?** It is unlikely that you're going to wipe any bed bugs on yourself when you get out of the shower. But don't leave towels sitting on the bed—that's a sure way to move undetected bugs around the room.

3. **Well, what about packing my own sheets from home? That has to matter, right?** If they are in the room, they're hanging out on the mattress, in the box spring, and behind the headboard. Besides, any reputable place will change the sheets every day anyway. And where does it stop? Are you gonna bring a comforter, pillow, and air mattress, too? Plus, it's nuts—why put your clean sheets on their potentially infested beds, and then take them home with you?

4. **But my aunt's house is spotless, so I'm safe, right?** Whether she has dirty dishes piled up

in the sink or floors so clean you could serve dinner on them, bed bugs could be waiting patiently for their next meal—you.

5. **This suite cost me $700 a night. Doesn't that buy me some protection?** Yeah. If it makes you feel better, we can call the bed bugs there extra fancy bed bugs.

6. **Okay, my friend lives deep in the 'burbs, with a lawn bigger than a city park. And she's never been outside the country. So her house isn't on bed bugs' radar, right?** There was a time when business travelers and their briefcases were the main source of domestic bed bugs, but those days are long over. Your best friend could live in the middle of nowhere and still have them in her guest bedroom. I once treated a guy who was so reclusive he made J. D. Salinger look like a social butterfly; he lived alone in a big house surrounded by hundreds of acres of land. It's important to remember: even recluses buy used furniture and books, and have the cable guy stop by every now and then—who knows who brought them into his home?

7. **It's a fairly new hotel—that has to count for something, right?** Sure; it has fairly new bed bugs!

## Step-by-Step: Hotel Room and Cruise Ship Cabin Inspection

There are some things in this book that you can choose to do or not do, depending on your living situation, lifestyle, or level of paranoia. This inspection is not one of them. You should do as much of the following as you can when you check into a hotel room or settle into your cruise ship cabin; pay special attention to things bed bugs love, such as wicker furniture (lots of places to hide) and popcorn ceilings and walls (ditto):

1. Put your stuff in the bathroom or on a table as far away from the bed or any upholstered furniture, drapes, and walls as possible.
2. Slip on your disposable gloves if you have them.
3. Pull out your powerful flashlight. Use it to check the following places, looking for blood (also known as spotting), fecal matter (dark spots that look like pepper), shed skins, and, of course, bed bugs (if you find spotting, try to smear it with a wet cloth; if it runs, bed bugs are likely present):
   - headboard (if it's bolted to the wall, check all around the seam where it meets the wall; they like to hide there)
   - bed frame, paying special attention to platform beds and those with drawers underneath
   - bed pillows and decorative ones
   - sheets; slowly pull them away from the mattress as you're inspecting

- mattress pad
- mattress; leave any encasements closed, but if the mattress isn't encased, pay special attention to both sides of the piping around the edges
- box spring (including the plastic corner guards)
- bed skirts and/or dust ruffles (both sides)
- baseboards, especially those closest to the bed and nightstand
- nightstands
- behind any artwork on the wall over the bed
- easy chair, couch, and any other upholstered furniture
- window and door frames
- draperies that touch the floor, bed, or nightstands
- wall/ceiling junction and crown moldings

4. If there is another bed, search that one, too.
5. If you brought it, spray Steri-Fab everywhere you searched. This is optional.
6. Breathe a sigh of relief!

## Step-by-Step: Private Home Inspection

When it comes to a friend's or a family member's home, it's pretty much the same search as what you do at a hotel, but it's more difficult. Think about it. Most hotel rooms have exactly what you need, nothing more; there are no dusty tchotchkes on a table in the corner, no stuffed elephant holding court on the bed, no bookcases overflowing with frayed volumes, and no closet full to bursting with abandoned exercise equipment and clothes no one has been able to fit into in a decade. Streamlined, well-edited spaces make for a much simpler search. Plus, in hotels, you're generally on and around the bed, so the bugs tend to hide nearby. But homes are a whole different animal. You can inspect, but with so many places to hide, you'll have a much harder time finding anything. But you must still do the hotel search, and also check any soft matter in the room, such as stuffed animals and extra blankets in the closet. Follow my tips for unpacking (hint: don't unpack), and don't skimp on your routine when you return home (see page 80 in chapter 3).

## What to Do if Your Room Is Infested

Uncovered a tiny bloodsucker while searching your room? Don't panic. First, whip out your cell phone or camera and snap a picture. It will give you proof of infestation and help with identification should you carry them back to your home and need to take legal action. (If you found drops of blood, or black "pepper stains"—bed bug poop—you can take pictures of those, too.) If you're at someone's home, you should tell them what you saw, and what you suspect it is, using the pictures as proof. I'd get the heck out of there (the longer your exposure, the greater your chances of picking up a tiny hitchhiker or two), but if you can't, see the next section.

If you're in a hotel, grab your stuff and head for the front desk. Ask to see the hotel manager and the head of housekeeping. Show them your photo (or tell them to check the room themselves if you weren't able to take a photo) and ask for another room. It should not be directly next to, above, or below your old room. (You can move to another hotel if you want, but good luck with finding another room at midnight.) Do not let them convince you to stay in your original room, and if they claim they can't move you, get a refund and leave. If you see signs of bed bugs in your new room, bolt for another hotel—the likelihood of picking up bed bugs is directly proportionate to the amount of time you and your belongings are exposed to the bugs. When you get to an Internet connection, you should contact the hotel's office to follow up, and contact the corporate headquarters to report what you saw and how it was handled, whether you were pleased or peeved. If the city you're staying in has a reporting service set up (like New York City's 311 program, the city health department, or an advocacy group's website), use that, too. Then head over to Bed BugRegistry.com and inform other intrepid travelers so they can stay far away from that hotel—you'll have good bed bug karma!

### Okay, But What if I Can't Leave?

Discovered bed bugs halfway through a sold-out week-long cruise? Or stuck at a friend's home in the middle of nowhere with no alternative place to sleep? Don't panic. If you're stuck in a cruise ship cabin and can't move, get a discount on your trip and ask housekeeping to vacuum your room (including all furniture) and spray judiciously with Steri-Fab if you have it. Live out of your suitcase and be sure to keep it (and all of its pockets) zippered closed (and your luggage liner or garbage bags sealed) to keep out bugs. Then drink heavily for the remainder of your trip! (Just kidding!)

If you can't leave your friend's home, move to another room. (Don't worry too much about drawing them into another room; that can take a day or two, and it's better than being bitten every night.) Keep your belongings encased in the plastic in your bags, hit your luggage with Steri-Fab as soon as you get outdoors, and carefully follow your returning home procedure (page 80).

## How to Unpack Safely

Okay, so you've done your search, and all looks clear. You're probably wondering what to do with your stuff. It's pretty simple: not much. At a friend's home, you want to stash your things as far away from the bed and the closet as possible. If the room has a large windowsill far from the beds, that's a good place to sit it. A table works, too. If you're in a hotel and those spots aren't feasible, you should put your luggage in the bathroom or on the top shelf in the closet or wardrobe. If you need to use a luggage rack, use caution—that's where everyone else puts *their* bags. Inspect it with your flashlight, then spray it with Steri-Fab if you have it.

Once you find a resting place for your bags, leave your stuff alone. Don't put it in the dresser drawers; it's safest inside your luggage encasements or sealed plastic bags. Keep luggage zippers and pockets closed at all times. If you have a jacket or some fancy dress you need to take out, hang it in the closet, preferably in a protective garment bag cover—don't lay it on the bed or the back of an upholstered chair. Reseal your bag each time you take something out, and you'll keep bed bugs out of your stuff.

## The Five Biggest Mistakes People Make When Sleeping Over

So even if your inspection comes up clean, there are some things you really shouldn't do when you're staying somewhere other than your own home. It's a cliché for a reason: it's better to be safe than sorry.

1. **Skip the inspection.** Seriously, you have to do a check, even if it's rudimentary, even if you just *know* your BFF doesn't have bed bugs.
2. **Put your suitcase (and everything in it) on the bed.** Anything that escaped your search could climb onto that toiletry bag and be deposited right in your suitcase when you put it back.
3. **Put your stuff on upholstered furniture.** Ditto.
4. **Place your clothes in drawers.** Even if you inspect them, the nooks and crannies make it hard to know for sure that there's nothing there. Don't take the

risk. If you must take clothes out of your bag, hang them in the bathroom or closet.

5. **Ignore bites.** Mosquitoes are probably not the source of those bites you got while in Detroit over Christmas break, and spiders are not a likely culprit in the northeast. If you're getting the itchies eight hours after sleeping in a foreign bed, don't get back in it.

## CHECKLIST: How to Make a Clean Getaway

If you did your room inspection and unpacked according to my guidelines, it's easy to make a graceful exit. Here's how to leave possible parasites behind:

- Rezip your luggage liners or tightly tie the plastic bags encasing your stuff.
- Zip your luggage.
- Spray the outside of your luggage with Steri-Fab to kill anything clinging to the outside. You can do this in your room at a hotel; if you're staying with a friend, you might want to wait until you get outside lest you offend your gracious host.

## Legally Speaking

Can you sue if you pick up bed bugs at a hotel? How about if you get them at a (perhaps ex-) friend's home? Just as with encountering bed bugs in other situations, the name of the game for establishing a successful case is negligence. That means you need to prove that the hotel (or your buddy) knew about the problem and did not act in a timely and reasonable manner to rectify it.

With a hotel or cruise ship, it might be easier to establish a claim than on a bus or at your cousin's home. Why? Because it's easier to locate the bed bugs and to corroborate your complaint with those of others who have stayed in the same room. Just be sure to give adequate notice. While you should definitely notify the front desk of problems, it's also important to file some form of written notice, because you'll need documentation to establish that a complaint was actually made. In a court case, your attorney will be able to obtain any documents filed. This is also a good time to make use of any formal electronic complaint process the hotel employs on its website, so you'll have your own copy of the record.

Private homes are just as difficult at modes of travel when it comes to proving negligence (see page 81 in chapter 3 for more on that). In any case, a successful claim could result in reimbursement for extermination and doctor's bills, as well as money for pain and suffering, and punitive damages if the situation was handled especially poorly (for example, if a hotel knew of a specific infestation and continued to sell the room without making any attempts at treatment).

## The Bottom Line

You can still get your beauty sleep in a bed that's not yours. You just have to be extra vigilant every step of the way. To review:

- Research is key; use the Internet to see if anyone else has had a run-in with bed bugs in the place you're considering (pages 88–90).
- They might lie, but it pays to ask hotels directly if they have treated for bed bugs (page 89).
- You *must* inspect your room before you settle in (pages 92–93).
- If you see signs of bed bugs, you have to report it and *move* (pages 94–95)!
- Avoid the mistakes everyone makes when sleeping over, and you'll be safer for it (pages 97–98).
- The best thing you can do to protect your home after staying in a hotel or at a friend's home is to follow my guidelines exactly (page 98).

# 5

# "We Like All the Same Places!"

Your Office and Eight Other Unexpected
Bed Bug Hangouts

At this point, you've read about bed bugs in your house. And the ones that are lurking in your car, on the train, in luggage holds, in airplane seats, and especially the ones hoping to make your acquaintance in hotels and motels around the world. What, you may be asking, could possibly be left? I'd love to tell you, "We've covered it, you're good to go," but, sad to say, you're not, because the epidemic has reached a tipping point. So, take a deep breath and get ready to read about a few more places—some very common and some, as the problems are becoming more common—where bed bugs lurk and where you need to take some extra steps.

And while theoretically they can be in any space thicker than a business card, from me to you, here are the key things you need to know to minimize exposure to bed begs in some of their (and your) favorite hangouts.

## The Office—And Bed Bugs Like to Put in Overtime

What do *Elle* magazine headquarters, Bill Clinton's famed Harlem offices, Rupert Murdoch's NewsCorp, and the Brooklyn District Attorney building have in common? Aside from the headline grabbers who have paraded through each, these spaces have all been visited by persistent, pesky bed bugs. And though their big names made the infestations newsworthy, offices and office buildings all over are battling the same problem.

### How Did They Get to My Office?

You already know that bed bugs don't need beds or sofas. They love people. And where do these beloved people spend most of their time? At work!

More important, your coworkers may have bed bugs at home. And at some point they can easily carry them to work in or on their clothing and belongings. Not to mention the countless people coming in and out of office buildings during the day, including consultants, workers, the mailman, messengers, and Mr. Lunch Delivery Guy.

The biggest problem with office bed bugs? Once a space is infested, your coworkers can continue to transport them from home to work, work to home, in an itchy, vicious, little bed bug cycle.

There's one other major hitch to clearing office

infestations. Office bed bugs don't cling to the obvious spaces (like beds in your home). Since people at work are generally moving, bed bugs adapt by hanging out in cubicle walls, books, files, in interoffice envelopes, chairs, and in picture frames. It means you can't do a check for blood or feces the way you can on your mattress. It also means a pest management professional can't as easily pinpoint where they'll be (which, again, is why it's necessary to call in a canine detective if you suspect your office has them).

Believe it or not, as much as it's disturbing to think of bed bugs taking up residence where you're trying to do your job, there is one advantage. In your office you have a much greater chance of spotting a bed bug on your desk or computer monitor because they're forced to come out to find food.

### Short of Quitting, How Can I Protect Myself?

1. **Put your stuff in a garbage bag when you get to work, and don't worry about where you're sitting.** You're much more likely to get it from your stuff that's in one place for a long time than from a chair.

2. **When you get to work, throw your coat (and your office sweater or shawl, unless you're wearing it) in the closet or on a hook at your cubicle, not on the back of your chair.** Nestled in the closet, it may still encounter your coworker's coat, but at least there are not people parading past it all day, leaning on

it with folders and other things that could have eggs or bugs on them. To be extra safe, the best method is to store it in a garbage bag and leave at your cubicle all day.

3. **When the job comes home with you in the form of papers, folders, and reports, you can protect yourself.** PackTite or vacuum the papers if possible. Even if you're not interested in attending to a hundred-page report with a vacuum, you should vacuum and Steri-Fab the area where you were working in your house.

### But How Can I Tell if My Workplace Is Infested, and What Can I Do About It?

Typically the only way you'll know is if you manifest bites or find actual bed bugs. Again, it is rare to find bloodstains, fecal matter, or bed bug carcasses. And if you do, that means you're already in the middle of a major infestation, and you need to ring the alarm bells.

### Just How Do I Ring an Alarm?

Now is not the time to be demure. When it comes to bed bugs, everyone needs to know what's happening. Sounds great, but just how do you do that and still have a job?

You may want to inquire if your company has a policy about bed bugs and what to do if they get into a workplace. Do they have protocols for alerting people in the area that there are bed bugs? Will they treat it appropriately? Does the company routinely bring in dogs to inspect? Do they have someone who's highly qualified to deal with bed bugs (and not just the contracted roach and mouse guy)? These are all things you want to know and, if possible, encourage your company to do.

And on the plus side, there is a trend for companies to implement a comprehensive strategy to deal with bed bugs in the workplace. Unfortunately, many companies don't take serious proactive measures until the second or third occurrence, incurring greater extermination costs and hours of lost worker productivity. My hope is that once they realize that the remediation costs are significantly higher than the preventive costs and engender much more employee goodwill, more companies will begin to elect this course of action.

### What if My Office Refuses to Do Anything?

The worst thing a major company can do is ignore employees and their concerns or not communicate effectively about the treatment process, including chemical safety and transmission from office to home. While diplomacy and tact are always the best options, if all else fails at getting your company to care, OSHA is there for you. If somebody calls the Occupational Safety and Health Administration (OSHA), a federal agency whose role is to insure safety on the job site, that's a game changer—and their fines start in the thousands of dollars.

### Who's Paying?

In a perfect world, if you had to get something as imperfect as bed bugs at work, you'd like to think that your office would pay for all costs associated with getting rid of them, right? And in certain offices, that may happen. But you are not legally entitled to payment, so don't bother shouting "I'll call a lawyer," when no one hands you a check for your troubles.

The good news, according to legal experts, is that you may qualify for workers' compensation if you suffer from bites you received in the office and you can prove that you got them at work and not somewhere else.

## Have a Seat: Bed Bugs at Movie Theaters, Playhouses, Restaurants, and Your Therapist's Couch

Yes, no one likes to think of these places as hot spots for bed bugs. These are the places where you date, are entertained, and get your mind right. But they are also the places where lots and lots of people sit without moving for long periods of time, on seats often made out of cushy cloth. In other words, as New York's Lincoln Center discovered, bed bug heaven.

Since you can't go spraying Steri-Fab all over the

place, and the Tyvek hazmat suit isn't quite fashion forward yet, here are some basic rules that will work: bags on your lap, never ever on the floor. Coats on your lap if you can, or the back of your chair. When sitting, try to choose slippery (from a bed bug's perspective) vinyl or leather chairs; if that's not possible, at least try to keep your belongings close to you. You can also purchase a clever new product called the Seat Defender, which acts like a condom for your movie or airplane seat.

## The Gym

Fortunately, bed bugs don't like the treadmill any more than you do. So there's not a high chance that you can get them on the equipment or the weights at your gym. Also, the free, fluffy towels that you love so much? Because they're laundered at such high temperatures, and because your wet or sweaty body is not attractive to them, you run little chance of rubbing bed bugs onto yourself with one. Whew.

So then, why are gyms on the Red Alert List? Because of these two little issues:

1. **The lockers.** Metal is better, wood is worse. Invest in a polyester (as opposed to cloth) bag; they're more protective. And throw it all in the dryer or PackTite as soon as you get home. Also, feel free to hit the outside of your bag with some Rest Easy for further protection.

2. **The mats and stretch/massage tables.** A few years
   ago, I would not have considered either of these
   areas, but today they are definite possibilities. You
   know the deal: gym clothes get laundered or Pack-
   Tited immediately.

## Shopping: To Buy Used or Not to Buy Used?

Any place selling previously owned clothing and furniture is at a higher risk for bed bugs. So vintage shopping needs to be done with greater care now. Not only
are vintage or thrift stores a problem, you should also
beware of consignment shops, the Salvation Army,
estate and yard sales, flea markets, furniture you find
on the street, and used bookstores (including used-
book vendors on the street), eBay, and Craigslist. Sure,
the price on that used dresser may feel like a steal—
right up until you factor in the cost of getting rid of
the bed bugs you bought along with it.

But before you give up vintage or consignment
stores completely, here are a few possible solutions:

- Regular clothes can be thrown in the dryer on high
  for thirty minutes. Dry-clean-only items should
  be dry cleaned immediately. And if you're lucky
  enough to own a PackTite, you run used books and
  other small items through it.
- In the case of furniture, you should vacuum, steam-
  clean, and Steri-Fab it (but be careful with wood)

the moment it comes into your home, but even then, there's always a risk with larger items.

### What About Used Items That You Didn't Buy but Borrowed?

When Shakespeare said, "Neither a borrower nor a lender be," I don't think he had bed bugs in mind. But since he said it, I suggest you heed his advice. In other words, take the precautions mentioned above when loaned items are returned or when you borrow something.

## The Mall and Department Stores

The problem, unfortunately, is no longer just used things, as proved by the reported problems in summer 2010 at Victoria's Secret, Abercrombie & Fitch, Hollister, and Niketown (not to mention scares at Bloomingdale's and Macy's). A few things to keep in mind about new items:

- If you try clothes on in a dressing room, take your clothes off because bed bugs are unlikely to attach themselves to your bare skin.
- And when you're trying something on, never throw your clothes on the floor. Put them alone (that is, do not mingle them with the store's clothes) on a hook. And when you get home, you know what to do with your garments (that's right—throw them in the dryer for thirty minutes on high).

**SOME (OKAY, TWO) THINGS YOU DON'T HAVE
TO FREAK OUT ABOUT**

1. **The mail.** Yes, theoretically paper that is sent from an infested office or home to your home can carry bed bugs. Yet, few cases are traced to the mail. Now, considering we are in the grip of a bed bug epidemic, this could change, so follow some basic rules and keep all mail off your sofa or bed or nightstand.

2. **Money.** Cash seems like a field day for a bug that loves people and paper. Yet it isn't a great host for them. So while I've yet to learn of an infestation through bills, I *have* been called in to treat banks where customers have carried in the bug and made an unwelcome deposit.

## The Bottom Line

- You cannot tell as a layperson if the places you go for entertainment, exercise, or health care are infested. What are you going to do, bring a dog to check your movie theater seat? So stop thinking you can spot bed bugs at the multiplex. I can't, you can't.
- That said, still go to the multiplex and everywhere else you want to go.
- Then, as soon as you get home, strip down, dry or PackTite your clothes, put on your PJs, and rest easy.

- Purses, briefcases, and laptop bags do not go on the floor. Even in your office at work.
- Coats and bags go on a chair when necessary and on a hook or coat rack if that is an option.
- Be supervigilant in the office, and put coats and bags in a garbage bag or a Rubbermaid container.
- Assume you are encountering bed bugs at some point and vacuum your home weekly. Ongoing use of a qualified dry steamer will also serve you well.

# PART TWO

# Treatment

Yes, You Can Get Rid of Bed Bugs!

# 6

# So You Have Them...
# Now What?

Are You Happy? No. Are You Powerless?
Absolutely Not!

It happened: you have bed bugs. Dry your tears, shout
one last "Why me?!" and let's start getting rid of them
*now*. This is why I'm here. (It's okay—you don't have
to pretend you bought the book for my scintillating
wit.) This chapter contains my wisdom on everything
you can do before the exterminators walk in the door,
while they're in your home, and after they leave: all
the steps to get bed bugs out of your life and where
they belong (in bed bug heaven).

Sometimes when I discover or confirm that bed
bugs are in someone's home, I'll say, "Now is the
time to start drinking or to break out the recreational
drugs!" But I'm not serious: while the bed bug epidemic is real and getting bigger every day—mostly
because of what people don't know—you *can* roll it
back out of your home and stop it at your door. So go

ahead and have a drink if you want to, but then roll your sleeves up and start making things better.

Now the first thing you need to do is read. "Did he say *read*?" Yes, I did, as in put your finger here: flip back to page 37, and read chapter 2! I'll wait here. (If you already read chapter 2, please continue!)

———•·•———

Back so soon? I hope you didn't skip anything! But here we are, and you've already taken a huge step in the battle: *you have knowledge*! Up to now, all the knowledge has been on the bed bug side. They may not have gone to good schools like you did, but they know everything they need to know about you— where you live, where you sleep, when you sleep, with whom—everything they need to know to fulfill their life's mission: stay alive, preferably with a full belly, be fruitful, and multiply. Which brings us to your second huge step in taking the war to the enemy: don't do anything!

I heard a Zen Buddhist saying that goes, "Don't just *do* something—*sit* there!" That's good advice to you right now, at least for the moment (so far I haven't seen any studies about the effectiveness of Zen meditation in bug extermination). Most people's first reaction to getting bed bugs is panic, followed by despair, followed immediately by a desperate desire to do…*something*! Don't give into that desire. I'm starting with a list of the top don'ts in fighting a bed bug attack.

## Before the exterminator arrives, do not...

- **...bomb or fog.** Bug bombs or foggers atomize pyrethrin, a natural repellent made from chrysanthemum flower. To imagine the effect, picture the Thanksgiving Day Parade; now picture some maniac lobbing a tear-gas canister into the middle of it. Are people standing around comfortably? "I say, Old Chap, this tear gas makes one rather uncomfortable, don't you think?" Bombs and foggers drive bed bugs from where they are—to someplace else, nearby! If you have a bed bug problem in one or two bedrooms, a bomb or fogger will spread it to the rest of your place; if they are in one apartment, they'll spread to more. In an office building, it'll spread to multiple offices. This is a recipe for disaster.

- **...spot spray.** Avoid any spray you can buy at one of the big chains online or a local hardware store. A partial treatment using over-the-counter products may kill the bugs you zap directly, but it will make the problem worse the same way bombs and foggers do. It will push them deeper into your furniture, your walls, or your home, or the apartment next door. If you actually see a bed bug in front of you, don't spray. Whip out the vacuum cleaner or your (wife's) Prada shoe—both are more effective, not to mention greener. Plus you'll get more value out of the shoe.

- **...throw out your mattress and/or box spring.** First of all, most people cannot carry a mattress without banging it everywhere and dropping bed bugs all over their home. Secondly, given the existence of bed bug–certified mattress covers, the only reasons for throwing out your bed are psychological (you were bitten in bed, so you convince yourself you'll be okay once the bed is gone) or aesthetic (you've been waiting for an excuse to get a new bed anyway). By the time it gets outside, half the bed bugs are in the building. Now someone sees a beautiful $900 mattress for free on the street—score! You think you got rid of the problem, but the guy who scored your infested mattress may be sitting next to you on the bus. Talk about karma!

*By the way:* In Boston the law requires that when you dispose of a bed-bugged mattress or box spring on the street you must put a special red warning sticker on it that says, "This may contain bed bugs." It's a

**IF YOU INSIST ON THROWING OUT YOUR BED...**

Getting rid of your bed when you have bed bugs is kind of like yelling at your car for getting you stuck in traffic, but if you are bound and determined to do it, there's a way to do it right. Wrap it in two to three layers of painter's tarp or—even easier—plastic wrap, seal it airtight, and keep it in your home until the day it's supposed to be picked up and carted away. (Wow, sounds like a lot of work— you sure you don't want to get a mattress cover?) There are also special mattress disposal bags you can buy that are like a giant sandwich bag, for about twelve dollars apiece.

CITY OF BOSTON

INSPECTIONAL SERVICES
DEPARTMENT

**CAUTION
THIS MAY CONTAIN
BED BUGS
<u>DO NOT REMOVE!</u>**

**CUIDADO
ESTO PUEDE
CONTENER
CHINCHES
<u>NO LO REMUEVAS!</u>**

MAYOR'S 24 HOURS SERVICE
617-635-4500

great idea, though I don't know how effective it has really been, or how much they enforce it...

Assuming you plan to replace the old mattress with a new one (unless you feel you'd be safer in a hammock), make sure to encase both the mattress *and* the box spring *before* bringing them into the house! Aside from the occasional scam of used mattresses being sold as new (just what you need), many mattress delivery trucks are infested, precisely because they often pick up the bug-filled mattress you're tossing out when they deliver your new one! Stores and companies often claim to treat their trucks regularly, but do not rely on their assurances—get the encasement on before the bed walks (or crawls) in your door! (And make sure the encasement doesn't get torn in transit.) Just so you know, I have never had a single company take me up on my offer to create a preventative bed bug protocol. You'll never be able to prove it came from their trucks. At the end of the day, though, most well-known companies don't want their name trashed—they'll pay the costs of the extermination, though they may never admit they were the source (or the conduit) of the problem.

## Under Penalty of Law...

You know the tag on your mattress that says "Under penalty of law this tag may not be removed"? People have been joking about that tag forever ("What happens? A signal goes to the FBI?"), or at least since 1913, when the law requiring it was passed. Before the law, unscrupulous mattress makers were taking old mattresses, giving them cosmetic cleanups, and selling them as if they were new (even though the mattresses were chock-full of stuff that would have belonged in medical waste sites, if they existed back then). As fearful as we are of the mattress police, I don't know too many people who have served serious time as a result of ripping off that tag in their own homes.

Since the law was put into effect, some mattress makers have *continued* taking old mattresses, giving them cosmetic cleanups, and selling them as if they were new. I know it's crazy, but it's true. In many states, including New York, it is completely legal to refurbish a mattress and resell it—but as refurbished, not as new! The law says you have to "sanitize" mattresses, but the law never defines the methods or protocols for doing that, and for a very good reason—there is *no way* to do it as I pointed out in the New York City Council hearings back in 2006. That makes it very easy for manufacturers to comply with the law.

People sometimes call me for my recommendation on where to buy mattresses. I don't have an answer, but I will tell you this: don't assume it's safest to get a mattress from a local place because it will come from right across the street. No matter how close the bed store is to your home, the mattress will be sent from a central warehouse.

## Can This Be Saved? (The Fate of Your Furniture)

There are a lot of stories about people losing their furniture because of bed bug infestation. Let me clear up a few misconceptions: most furniture can be treated by your exterminator. Using techniques like dry steam, vacuuming, and carefully applying pesticides in cracks and crevices when needed, chairs, desks, tables, and sometimes even couches (one of the most difficult items to treat) can be saved. And if your wallet can afford it, you also have the option of fumigating the contents of your home and not losing a thing.

Okay, hopefully I've stopped you from bombing, fogging, spot spraying, or throwing out your mattress and furniture—all of them great ways of tak-

### THE BEST WAYS *NOT* TO FIGHT BED BUGS...AND THE REASONS WHY

When it comes to bed bugs, do not listen to your gut—your gut knows nothing about getting rid of bed bugs. Nothing your intuition tells you to do will help you, and much of it could actually hurt you and other people. Don't:

1. **throw out the mattress.** Not only do you stand a good chance of leaving a trail of bed bugs

and eggs as you go, but you'd be surprised how often people pick up things from the curb, thinking they've scored. Don't be the Typhoid Mary of the bed bug set.

2. **throw out all the furniture.** It's a waste of money—a good strategy can save all your stuff—and a great way to spread these varmints to your neighbors.

3. **wait and see.** And while you wait, a small problem will turn into a major (and majorly expensive) one.

4. **sleep on the couch.** If someone took your plate and put it on the dining room table, you'd just mosey on over there and grab it. You = a bed bug's plate. Move to the couch, and I guarantee they will follow you.

5. **head to Home Depot.** Here's some news: those sprays they sell in DIY stores across the nation are just a step above mouthwash in strength and effectiveness. Seriously. They work only on direct contact and will just force bed bugs to find new hiding places in your home.

6. **use your mother's steamer.** Unlike recommended dry steam vapor cleaners, these things are not hot *or* dry enough to kill bed bugs. All you'll do is warm the bugs and blow them around. Wheeeee!

7. **blast them with the blow dryer.** You're just coiffing the bed bug's hair follicles and giving it a free ride across the room.

8. **just say, "That's it—we're moving."** And the bed bugs are getting new digs!

ing you from frying pan to fire. (For more *un*tips, see the box on page 122, "The Best Ways *Not* to Fight Bed Bugs.") But there are some things you can do on your own before the dog comes in and the exterminators arrive, things that will actually improve your situation.

And, by the way, when you're bringing affected clothing from bedroom to washer/dryer, don't carry it against your chest. Put it in sealed plastic bags or, better yet, dissolvable laundry bags, and *then* bring those bags to the washer, wash on hot and place in dryer on hot for at least thirty minutes!

Here is how people in apartments and dormitories often spread the wealth. A guy calls and tells me he has bed bugs. Before I make an appointment, I run through the whole prep list with him, including telling him to wash and bag all clothes, etc. Completely disregarding the part where I say, "Bag everything tightly before you wash it," he takes his stuff down to the laundry room, and puts it on the table, dropping eggs, mature bugs, and everything in between. He washes his clothes, bags them up, and leaves, happy with himself. Then the next person comes down, puts *her* stuff down on the table and—you guessed it—picks up the bugs Mr. Share-and-Share-Alike has so generously left behind.

## Do...

1. **...encase your pillow, mattress, and box spring right away.** By encasing them, you prevent any bed bugs inside them from getting out. ("What? *Keep* my bed bugs?!") You probably don't like the notion of sleeping on three millimeters of cloth with bed bugs underneath, but it works. Bed bugs trapped inside encasements cannot get out and will die of starvation. It will keep others out and force them into other areas that will be treated and more observable. With bed bug–certified encasements for your pillow, mattress, and box spring (see chapter 2, pages 40–42, for details), you will not have to worry that you are being assaulted from *within* your bed. And make sure you don't tear it—encasements are good as long as they're intact, but if they tear, they're no good. Use cloth or caulking to cover anything sharp or protruding in your bed frame and headboard, and make poking holes in your encasements less likely.

2. **...throw it in the dryer.** Bed bugs can't take the heat—much as they like your sheets, they don't like being washed or dried. Their optimal temperature range is 70–80°; temperatures of 130° and up for at least half an hour will kill them. Setting your washing machine on hot *should* be enough, but in many washers the water does not get hot as

it should—hot water by definition leaves your hot water heater at 130°; it arrives at your washer some ten degrees colder and can lose heat from there. Who wants their bed bugs freshly bathed and out for vengeance? Your clothes dryer beats that by at least fifty degrees, which is plenty to send bed bugs on to the world to come. So put any clothes, linen, bedding, drapes, curtains, etc., that can handle the dryer, in the dryer—that includes dry-cleanables, which can't be washed but *can* be dried. Almost every source says half an hour (or even less) will do it, but I tend to give it at least thirty minutes just to make absolutely sure.

3. **...bag it.** When the dryer is done take out all your items, put them *immediately* into very sturdy and clean plastic bags, and seal them.

4. **...bake whatever can't go in the dryer in the Pack-Tite.** In case you didn't go back and read chapter 2 (despite my specific instructions!), the PackTite is a portable heating chamber, a big nylon frame (the size of an old-fashioned suitcase when folded, or a midsize dog crate unfolded) with a metal stand and a small intense heater on the bottom. You can use it to bake the life out of any bed bug foolish enough to hitch a ride on your possessions. It heats up to about 130 degrees, where you should keep it for an hour, but don't be in a hurry—the more you have in the PackTite, and the more densely packed, the longer it takes to reach optimum bug-baking levels.

These Do items are general rules. For specifics, see "Prep Your Place" on page 129.

Here are *the Big 3*—the stop-drop-roll-and-do-*now* action steps for tackling the problem before it gets worse.

## First: Call in the Dogs!

The bed bug–sniffing dogs, that is. Once you suspect you have them from home self-inspection, get the dog to confirm it, because reputable exterminators will not treat without confirmation. With the rise of the epidemic, more and more dogs are being trained specifically to find bed bugs wherever they may hide. Seriously! Think of them like those dogs trained to sniff out cocaine (which always seems a little dangerous to me—one sniff too much and they're hooked), or truffle pigs, except nobody wants to eat what these pooches snuffle out. Training a dog is serious business

### GOING TO THE DOGS? WHAT TO KNOW BEFORE YOU GO

1. **Is the inspector shilling for a pest-control operator?** To avoid conflicts of interest, a bed bug–sniffing dog and his trainer should *not* be owned by an exterminating company. Otherwise you could have someone finding (or

worse, planting) a problem. ("Oh, wow, you're totally overrun. You'd better call ABC BugKillers—they're the only ones equipped to handle an infestation like this!") You should feel free to ask if there's an exterminator they recommend, but be wary if it seems like they're trying to push a particular company on you. Independent dog companies have no vested interest—whether they find bugs or not they get paid the same. They have one product, one service, they keep the dog well trained, and that's it. If an exterminating company has an in-house dog, they may be doing too many other things—the dog could be tired, its training may be lax—and of course there is a big payoff for them to find bugs in your home!

2.  **Is the inspector "moonlighting" as an exterminator?** Sometimes an unscrupulous dog inspector will find a problem, then tell you it's not too bad and, hey, by coincidence, he just happens to have a tank of the best pesticide out in the van! Next he'll tell you what a PCO would charge, and offer to do it for half of that price, right now! This is a classic case of sounds-too-good-to-be-true-because-it-is; so watch out! As I say, you stand your best chance when the dog is from an independent company.

(it costs approximately $12,000 to buy a trained dog), and their trainer inspectors are serious about it, but there are things to look out for—see the box "Going to the Dogs?" pages 127–128.

## Second: Prep Your Place!

Any reputable company will have you scrupulously prepare your home in advance so it can be treated properly—if they don't, the company is not qualified and not thorough in its approach. There's a huge learning curve for each and every technician. A lot of small companies don't have the resources to train their technicians and keep them up to date. Some companies, like mine, won't even schedule an appointment unless you have fully prepared your home. So once the dog inspection has carefully identified specifically which rooms are "bugged," here is what you have to do, step-by-step:

- Bedrooms: Strip all blankets, sheets, pillow cases, bed pad, dust ruffles, etc., off the bed, and run them through the dryer at the hottest setting for half an hour; the same goes for *all* clothing in dressers, drawers, closets, hope chests, etc., and any fabric *on* your bedroom furniture as well. As soon as each batch is done, put it in bags and seal it. Then, if you haven't done it already, put your mattress, box spring, and pillows in bed bug–safe encasements. (See the "Do" list on pages 125–126.)

129

- Anything that cannot or should not go in the dryer (shoes, suits, coats, etc.) goes into the PackTite. (See the "Do" list on pages 125–126.)
- You can use Steri-Fab on all furniture surfaces (be careful with wood surfaces or leather because they may discolor).
- Empty closets, bedside tables, desks, and dresser drawers for treatment. Place items in clean plastic bags and tightly knot the bags. Do the same for items like handbags, luggage, shoes, etc.
- Put all papers and books in plastic storage bins or garbage bags. Unless you have a PackTite, I recommend that these items stay sealed up for nine months. If you have to get to some of them sooner, keep them in a separate bin and talk to your exterminator about fumigation or other alternatives, like the PackTite. (See chapter 7.)
- Bed bugs abhor a vacuum (cleaner). Use a powerful canister vacuum cleaner with a bag that can be discarded after each use (remember, bagless vacuums are a big no-no!) to vacuum every crack and crevice in every affected room, starting from the bed and working outward. For starters, vacuum any bed bugs or eggs you see. Vacuuming *abrasively* (see the box "Bed Bug–Vacuuming Technique") will extract *all* stages of bed bugs, even eggs (which can be tricky because eggs have a kind of glue that makes them stick to whatever surface they're on). This technique works to dislodge bugs and eggs, making them easier to suck up.

**BED BUG–VACUUMING TECHNIQUE**

Use a crevice tool anywhere you can, and be as abrasive as possible in scraping the tool along surfaces to dislodge bed bugs and eggs. In the bedroom, work from the bed outward; go from the headboard to upholstered furniture, and on from there. When you're done vacuuming, *do not* leave the vacuum in the room! Throw out the bag after use; any time you stop vacuuming for more than twenty minutes, seal the bag and throw it out. Don't use brushes—bugs can get caught in the bristles. Vacuuming alone will not solve a bed bug problem, but it is one of many things that will help bring down the level of infestation, and it's green, immediate, and effective!

- Make sure the technician has access to all rooms and closets to be sure that an inspection and treatment can be done properly. That means clearing the floor area as much as possible before the exterminator arrives, especially in rooms where you know or suspect there are bed bugs. The perimeter(s) of the room(s) must be accessible for the technician to conduct an inspection and treatment services.
- Take the bed frame apart for the treatment. The mattress and/or box spring should be removed.

- Take down from the walls all pictures, clocks, posters, hangings, and any other wall décor, *including* shelves. Vacuum them and seal them in plastic.
- Light switches and electrical outlets are common bed bug hangouts. Loosen all switch plates and outlet covers so your pest control operator can properly treat them.
- Wall-to-wall carpeting: a classic textbook approach would have you peel back the carpeting to treat it. That may be the case in particularly horrible infestations, but I have done thousands of jobs where the carpet stayed where it belongs. If you're not overrun and a pest control operator tells you before walking in the door that your carpet will *have* to come up, get a second (or third) opinion.
- Remove all your pets from the premises, or at the very least *move* them to a part of your home that is not being treated.
- Safeguard your valuables (seal them away, too!), and remove all breakables from all areas that are going to be treated.

## Third: Call in the Professionals!

The moment dog inspection has specifically identified all the infested rooms in your home (within reasonable limits—dogs can't sniff what is too far above their noses), it's time to get yourself a pest control operator (PCO). In the next chapter you'll learn how to find a PCO who's actually qualified to exterminate bed bugs, but for now I'll just say:

- **Ask how much bed bug work they do.** How often, how recently, and since when.
- **What are their methods?** The best answer is actually, "It depends." Be wary of someone who tells you, "We use Cryonite—it's the best!" or "Thermal remediation is the only 100 percent–effective approach!" The reality is that the best strategy depends on a lot of factors (Apartment or house? Small area or large? Caught it just in time or total infestation? Do you have somewhere else to stay...? Are you on a tight budget?), and often it is a combination of approaches.
- **Can you call references?** Naturally companies will refer you to their most satisfied customers, but people are usually pretty honest about the service they've gotten—and nobody who's had bed bugs wishes them on anybody else! Any quality company should have commercial accounts, such as landlords or managers of large buildings, and clients who will level with you.

This is not a simple process, or a short one. Most of us are pack rats—even when we think our homes are not cluttered (compared to *some* people we won't mention), we have a lot of stuff! Preparing your home for bed bug extermination has almost all the aggravation of moving, with none of the payoff of moving into a new home. Just like moving, most people have to take off from work to properly prep their homes. It takes the average person two to five days to prep—full-time!

If you can do it yourself, great—you'll save some money, and get a little satisfaction that you were in the front lines of the battle against the mite-size menace. Many people are not physically up to it; even if you can handle it physically, it might be more *emotional* stress than you can deal with right now. Folks who can afford it hire outside help—in some cities pest control operators offer prep services, sending well-trained people to ready your home before the actual extermination. The cost can go from around $50 an hour to way, way up (I know one company that charges $700 for four hours, then an hourly rate after that). Be *careful* of fly-by-night bed bug–prep services; they may do a superficial job because they are just not properly trained. If you have to go that route, make sure you're there, with this book in hand! ("Excuse me, Jeff said that pictures should be taken off the walls and vacuumed thoroughly.")

And be careful that the independent prep service is not offering more than it can responsibly take on—like steaming all the clothes and items it removes from your closets. Steaming is great for cracks and crevices, but your clothes can never be steamed thoroughly enough for you to know that all the bed bugs will be gotten—it's a waste of your time and money.

## JUST HOW BAD IS IT?

People rarely are aware of how bad the problem is. The earlier you treat, the better, of course, but it can be hard to gauge just how serious an infestation is. The natural inclination is to think you have taken care of it because you got it in the early stages and don't have bites. But unless your space is tiny and you really caught it with the first bug, it can be hard to localize it with absolute certainty. That's why no qualified PCO will guarantee that you are bed bug free after just one visit.

Sometimes I'm asked to quantify the severity of an infestation ("On a 1-to-10 scale, where do you think I am?") I don't know what it's worth to tell people "Well, the Smiths were in much worse shape but the Joneses pounced on it a lot faster than you did," but I do explain the basic strategy, which applies whether you live in a studio apartment or a mansion: once you have a single bed bug, we treat the problem the same as if you had a lot. It's the only way to solve the problem. In my company, whenever we hear that a place is "almost completely bed bug free" or "virtually bed bug free," we have this little exchange:

What do you call a bed bug job that's 99.99 percent effective?
A failure.

Bed bugs are like cancer, in the sense that if you don't get rid of it entirely, it will keep growing. You always have to treat the affected area *and* go beyond it. The more thoroughly you prep your place, giving us the greatest possible access to potentially infested spots, the less you will see of us. Your best shot at getting rid of bed bugs fast is cooperating fully with your exterminator by de-cluttering, cleaning, and providing access to all baseboards, electrical sockets, closets, and carpet edges, and moving all furniture that can be moved away from the walls.

If you think the prep is more than you can handle, or you don't feel you can take the time off work, or you've read stories online about people who got rid of their bed bugs with baby powder and alcohol—think again. I can't tell you how many people come to me after months or, worse, years of battling bed bugs on their own or with a host of incompetent exterminators. The only way to beat these creatures is with the right prep work followed by the right professional help. The more meticulous your prep, the better your chance of getting rid of them quickly and completely. The less careful your prep, the more likely you are to need costly and time-consuming follow-ups.

## IT'S ALL RIGHT TO CRY

Besides horrible itching, scratching, and other basic quality-of-life issues, getting bed bugs can feel overwhelming and take a real psychological toll on you. If it makes you feel crazy, I have news for you: you're not crazy! But that doesn't mean you're not suffering. Along with insomnia, a surprising number of people whose lives were invaded by bed bugs have bouts of what's called delusional parasitosis: you think you're getting bites even when you're not! (As if the actual bites weren't bad enough, sufferers are having their worst fears validated, in the worst possible way.) The anxieties that come as a dubious fringe benefit of bed bug infestation can last for months after the infestation is over. There are therapists who regularly treat people who have suffered (or are still suffering) from infestations, but any good therapist should be able to help you get through this malady that has been around a long time—since well before the current bed bug "renaissance." You're already doing everything you can to banish the dastardly little vermin from your home—don't hesitate to get them out of your head!

## Monitoring: What It Means and How to Do It

Let's say you've had the blasted things, you've done all the right things, the pest control operator has done all the right things, and the bed bug beagle has given you the all-clear sniff four weeks after your first treatment. Hurray! Are you home free? Can you go back to business as usual? Not unless you want to go back to bed bugs as usual!

Luckily, there is a middle ground between sleepless itchy nights and living in a dream world where bed bugs don't exist because you don't want to think about them. It's this simple: In addition to following the prevention protocol outlined in chapter 2, vacuum at least once a week or more for the next six to eight months (floors, headboards, baseboards, and every conceivable crack and crevice—in the infested rooms, of course, but also throughout your whole house, since you travel from room to room). This will reduce your chances of reinfestation from the original source if you haven't identified that source and prevent reinfestation from someone who may have recently gotten them from you. Don't forget to take the bag out of the vacuum, put it in a plastic bag, seal the bag, and get rid of it immediately.

I know this sounds like a lot. ("Vacuum every square inch of my home every week for six months? When am I supposed to eat, sleep, work?") It *is* a lot of work, but it's not really every square inch—and the more you do it, the faster (and more efficient) you'll get. Once you've survived the scourge, you may be surprised to find out that fighting its recurrence is not that hard.

So when do you know the infestation is over? The longer you go without bites and evidence of bugs, the safer you are. I generally tell people that they should be clear after a minimum of three or four weeks without bites. However, this can be complicated (as I've said previously) if you have a pet. In terms of your stuff, I recommend unwrapping it, little by little, day by week, over the next three to five weeks, just to make sure you're totally clear. And bring the dog in to check thirty days after your last treatment. Here's why:

- You bring the dog in, and it clears your home. Hurray—unwrap all the bags, right? *Wrong!* If the original source (nanny, work, your child's play date partner, etc.) is still affected, you can get them again. Or you may get it back from someone *you* gave it to! (As John Lennon sang, "Instant karma's gonna get you...")
- When it comes to your sealed bags you'll want to bear in mind that opening them too soon may either (a) re-expose your home because you accidentally opened a bag of untreated items (untreated items must remain bagged for at least nine months) or (b) expose bed bug–free items in the bag to potential reinfestation if your home is not completely clear.

If you unpack slowly—say, three bags a week—you'll know sooner if you have bed bugs in your bagged stuff, and prepping for the next treatment will be much less challenging. Better to unpack slowly and have less to do. If you *do* see them, you're still under warranty, and you may have only two or three bags to wash or treat.

## When Will the Paranoia End?

The experience of having bed bugs in your home can leave you a little paranoid—you may strip down to scrutinize yourself for bites, be hyperaware of marks on other people's bodies, and even change your thinking about how you want your home furnished. You may find yourself getting rid of your platform bed or captain's bed (a mattress sitting on a piece of wood is the ideal home for bed bugs), likewise bed skirts, upholstered headboards with fancy folds or pleats, popcorn walls or ceilings, wall-to-wall carpeting, floor-length drapes, or anything wicker (almost impossible to inspect or treat properly). You may opt for plastic or metal bed frames over wood. Whatever measures you choose to take, remember that the paranoia should not last more than a year, but good practices will protect you for a lifetime.

## Where Did They Come from, Anyway?

Most people are never sure where they got them. My vast experience has shown me that your likelihood of exposure is greater if you're single, if you have boy/

girlfriends, if you have roommates, if you have kids, if you travel regularly for work or pleasure, if you have people working in your home, if you take public transportation or taxis...If you didn't fall into *any* of those categories, I'm not sure whether to congratulate you or console you.

It's helpful to know where you got them so you can possibly avoid getting them from that source in the future, but my goal is to get you thinking, not blaming. You should not stop living your life because bed bugs are in the world. If you went out to eat before, if you went to the movies, to the theater, to concerts, to the library, don't stop! Just be aware of the possibilities and, above all, make your front door the line in the sand.

---

**BUGPEN: ONE OF THE DANGERS OF APARTMENT LIFE**

I got a call from the resident manager of a posh building on the Upper East Side of Manhattan. The grandson of one of his tenants, who came by every now and then to visit his grandfather, had set glue traps, and the traps were full, the manager thought, with bed bugs.

"Peter," I said. "That's impossible. Do you know how infested a place would have to be to fill up a glue trap? The walls would have to be crawling with them!" I told him to go up to the apartment, check it out, and call me back. When he did, I heard a stream of profanity to make a sailor blush.

"Holy f@#%ing s*&#! Son of a f@#%ing b#*$%!" and on and on and on. "I can't even believe it. The apartment is *moving*." I might add that this was the first time I'd heard this mild-mannered man swear in thirteen years. The amazing thing was how no other apartment was affected, but it turned out that the tenant was pretty much a recluse—everything was delivered, he left the place every ten weeks or so to go to the doctor, and that was it.

Remember Pigpen from *Peanuts*, with the cloud of dust swirling around him wherever he went? This man was Pigpen, only with bed bugs—they were not just in his living space, but literally all over him. The sad thing is that, as crazy as this sounds, he is not that uncommon. There are "Bugpens" everywhere, usually men who live by themselves, and not all of them are hermits. They're walking around, shopping, going to work, the gym, wherever, and all you need is to stand next to one of them for one of his puny passengers to take a leap toward greener pastures...

The worst situations are in the homes of elderly people. They have so many other concerns—they're too sick, too weak, their vision is poor, they have barely enough money to get by—and they may not have people visiting them regularly who would notice the problem. I can't tell you how many times the root problem is in the homes of people who are caring for them. Unfortunately, the elderly usually don't report infestations but receive help after someone else becomes aware of the problem.

There is no silver bullet. Fighting bed bugs is like trying to lose weight. You can pop all the miracle diet pills or buy all the instant in-shape machines you like, but at the end of the day, slimming down and staying

## HOME REMEDIES AND ECO-FRIENDLY SOLUTIONS: WHAT HELPS, WHAT WORKS, AND WHAT DOES NO GOOD AT ALL

There are hundreds of products on the market making all kinds of claims. "Why call an exterminator when Product X will do a hundred times more for a fraction of the price?" My question would be, "If Product X works so well, why do extermination companies with superior training, high-tech equipment, and access to restricted pesticides still continue to struggle with these scourges? Why does the hospitality industry spend millions a year fighting infestations?" I have been battling bed bugs for fourteen years, putting all my resources into updating and trying to find new ways to push back the waves of bed bugs. Chances are not great that you'll be able to do it on your own with a can of spray and a YouTube video.

trim is about lifestyle change, and it takes work. Most products are taking advantage of a hysterical market—when we are freaking out at 3:00 a.m. and can't talk to anyone, we are much more inclined to believe wild claims and the products they promote.

But beware—some products are not effective unless you apply them directly to bed bugs (in which case you might as well blast them with hairspray). Some online companies are selling restricted-use pesticides, which should be used only by licensed companies, to unlicensed people, sometimes with terrible results. In the best case these products will do nothing; in the worst case, they will make things worse—endangering you and your family's health. Most pesticide poisonings in the United States occur when do-it-yourselfers use these materials on their own.

## Commonly Sold Products That Can Work—Depending on How You Use Them!

Unless you catch a single bed bug strolling in your front door, spray it directly, and give it a proper burial, the products listed below are really effective only when used as a *supplement* to real treatment. [Caution: Always read the labels carefully and follow the directions precisely!]

• **Steri-Fab** is a disinfectant insecticide (pesticide, moldicide, fungicide) spray that I use on an ongoing basis in my home and office. I strongly recommend it. It kills bed bugs on contact (but not their eggs). According to the material safety data sheet (a standard form detailing all the properties of a given product), it has zero carcinogenicity. You can use it around your headboard, on the bed frame itself, and on any smooth surface, as well as on mattresses, rugs, carpets, and upholstered furniture. Other than kitchen utensils and waxed surfaces, you can spray it on pretty much anything that doesn't breathe. The health threat level for this product is minimal.

- **D-Force:** Its active ingredient is deltamethrin, a neurotoxin—so it hits bed bugs in their nervous systems, but for us it metabolizes and loses toxicity over time. It has some residual effectiveness up to eight weeks after spraying.
- **Bedlam** kills on contact and provides residual control, and is the only product that kills bed bug eggs. The active ingredient is sumithrin; also a neurotoxin, it's not supposed to be sprayed on people or animals.
- **Permacide P-1:** its active ingredient is permethrin. A neurotoxin like deltamethrin and sumithrin, it is toxic to cats as well as fish. The EPA classified it as "likely to be carcinogenic to humans," so even though it is available to the general public, it should be used very carefully.
- **Diatomaceous earth** is a nonsynthetic mineral dust that is often cited as an "all-natural" pesticide. My

colleague, Wayne Tusa, an environmental consultant, and the head of Environmental Risk and Loss Control, Inc., is wary of the "all-natural" hype. He says, "The dust and its silica content pose a respiratory hazard to the occupants. To control bed bugs, the product must be applied in immediate proximity to frequently occupied spaces. Because people move around, the flow of indoor air, and the product's low density, the dust will disperse resulting in occupant exposures. What the relative risk to that exposure is I doubt anyone knows, but environmental common sense would lead me to apply it only in unoccupied or enclosed spaces." I'm concerned that this unrealistic idea of its "greenness" is leading people to use it indiscriminately. And I advise leaving the use of this product to a professional.

The number one question I get, seventy-five times a day, is "Can you get rid of my problem?" I'm happy to say that **there has never been a bed bug problem we couldn't solve**—but it can take time. Ninety percent of our cases are completely solved within four to six weeks, from the time we walk in the door to the final all-clear. The more difficult cases take longer because there is more detective work involved, but through deeper digging we ultimately find the answer—whether it was the neighbor, the playmate, the mistress, or the butler who did it. A competent pest control operator is always part Sherlock Holmes: We never just walk in and start spraying; through detection, strategy, and cunning, we always vanquish the enemy. You may feel overwhelmed and even despairing, but just remember: you bed bug problem *can* be solved if you give it the time and focus it needs.

The best thing you can do right now is turn the page to chapter 7 and start reading.

# 7

# How Do I Find the Right Exterminator?

## And What Do I Do Then?

If the previous chapter was the first one you read, this will definitely be the second. You have bed bugs and you want to get rid of them. You are bracing yourself to do all the things I told you to do in the previous chapter, but first you have to pick up the phone.

## Who Ya Gonna Call?

Since Bill Murray and his pals are busy with other projects, you're going to have to do a little research. The easiest way may be word of mouth: if your friend had bed bugs, brought in Company X, and hasn't had a problem for a year, then great—by all means, call Company X. But it isn't usually that simple, and once you start getting into "I heard from a friend of a guy who knows this guy whose friend knew a guy...," you might as well be opening the Yellow Pages, closing your eyes, and calling in whomever your finger lands on.

The Yellow Pages is not the golden answer, but it's not a bad place to start. Call around, starting with the ones that highlight their bed bug work. Ask questions (See the box "What Do I Ask?" on pages 151–153.) Compare their answers. I know you want to get rid of your bugs as fast as possible, preferably yesterday, but don't just go with AAA Bed Bug Killer, Inc., because they were the first one listed and you're desperate. And don't make *The Biggest Mistake*: choosing the guy who gives you the best price, the guy who'll get there quickest, or the guy who does all your other exterminating.

## Who You *Oughtta* Call

Bed bugs are not like other pests; they require pest control operators who are trained and experienced in treating them. There are plenty of competent, decent, garden-variety pest control operators out there, but there are too few who are seriously trained in getting rid of bed bugs, let alone seriously experienced at it. If you are not very picky in finding the right one, you *will* end up with a worse infestation than the one you have now; it will cost you more money than you're already spending, give you more aggravation than you already have, and possibly endanger your mental health and the mental health of people you love.

I've had plenty of people come to me and then choose another company because it was less expensive, only to come back to me months later. Many companies will charge $475 to treat an apartment, but they'll also charge $175 for each return visit, and often they'll tell you they have to come back every week or two for months! Experienced companies, like mine, charge anywhere from $800 and up, but we provide a full warranty.

One couple knew they had bed bugs, but only the wife was getting bitten, so the husband didn't take it seriously. The company they went with was $400 cheaper than mine but, unlike us, didn't give a warranty—they just kept charging for every visit. And there were a lot of visits, because the exterminator did not have a vested interest in solving the problem and unknowingly pushed the bed bugs back into the wall *and* the neighboring apartments. Their co-op was not happy! By the time they finally brought me in to deal with it correctly, it ended up costing tens of thousands of dollars more than I had estimated, because dozens of apartments were infested. Sometimes the price of a given pest control operator is higher because of their financial investment in training and their skill and experience. If you find a good one, don't choose low cost over reputability.

**THE FIVE CRAZIEST THINGS I'VE SEEN CLIENTS DO TO PROTECT THEMSELVES FROM BED BUGS**

1. **Fog their bedrooms every night.** It doesn't kill them (they aren't roaches) and you'll just drive

them deeper undercover where they can lie in wait until their next meal.

2. **Spray themselves with pesticide.** You, meet acute toxicity poisoning.

3. **Wear a necklace made of garlic cloves on the advice of their grandmother.** Bed bugs might *act* like tiny little vampires, but the pungent smell of these herbs does nothing to curb their appetites.

4. **Pour kerosene in and around their beds.** The trifecta: smelly, ineffectual, *and* incredibly dangerous!

5. **Let a bunch of geckos run loose in the apartment.** No, seriously. We had a client who did this as her "green" solution to an infestation. The result? More bed bugs plus multiplying lizards. Her apartment looked like a Geico commercial.

## WHAT DO I ASK?

You can add to this if you like, but here's a list of the questions you *must* ask any exterminator before you give them your business:

• Is your company a member of the National Pest Management Association (NPMA)? Serious companies who are dedicated to their profession and moving the industry forward will always be a part of this group.

- Do you treat bed bug problems regularly?
- For how long have you been in business? (Ten years minimum in the pest control business—the *company* has to have been treating for bed bugs for at least seven years. A guy who worked for someone else and went out on his own two years ago is *not* an experienced detective and strategist. No matter how closely he worked with his previous boss. As the saying goes, sitting in first class does not make you a pilot.)
- How many technicians do you have? (Should be at least eight. Larger companies—but not too large—are better able to train, invest back into the company, and have more effective management. And because they have more to lose, they have more incentive to do things right.)
- What methods do you use?
- Do you warranty your work? (Once you treat, will you keep coming back for the same price?)
- How many technicians are actually licensed in the company, and is the person coming to my home licensed? (FYI: Some companies have only one licensed technician and the rest of the staff work under that person's license.)
- Are you accredited by the Better Business Bureau? What is your rating?
- (optional) Do you have nonchemical (i.e., green) options?

- They should also be asking questions, too, so you should also judge them by what *they* ask *you*—are there children in your home, asthma sufferers, elderly people...?
- Do you have references? (If possible, get references from property managers.)

An alternative to the Yellow Pages is the Better Business Bureau. Go to their website (www.bbb.org) to either check out a particular company (their name, phone number, website, and/or e-mail) or search "Pest Control Services" (if you don't have someone in mind). BBB accreditation is not a guarantee that a company will do the best bed bug exterminating, but it does mean a commitment to basic integrity and accountability. So don't go with a pest control operator only because it is BBB accredited, but try to avoid a company that *isn't* accredited. If you live far from an urban center you may not have as many choices, so rely on the questions in the box "What Do I Ask?"

## Beware of the Mega-Exterminators!

Be careful with exterminators who spend lots of money on advertising. They're not necessarily the best, and part of the cost to *you* will be paying for *their* advertising dollars. And a big company with dozens or hundreds of technicians may not prioritize you to the same degree that they do one of their major corporate clients. Plus, when there's a huge chain of command, you can get lost in the shuffle. If you have a problem that requires more than a simple solution, it can be difficult to get to someone who has a vested interest in solving your problem. But when you go with a smaller company and things aren't going well, you should be able to get the owner on the phone. It's hard to get that level of service with very large companies.

## Think Small—But Not *Too* Small!

So while you should be leery of mega-exterminators, you should also steer clear of companies at the other end of the yardstick. Little places, even with the best intentions, lose out on scale. They have a small back office (if they have one at all), which means you can't get the support you need even before they walk in your door, let alone after they leave. They don't have good training programs, which means their pest control operators are not up to date on technologies or techniques. If you call during normal business hours and get an answering machine, watch out. If you get a call back the next day from someone on the road or in the field, the "company" could be working out of a garage or a home. And if the main number is a cell phone, forget about it! All these are typical signs of an operation too small to take care of your problem in the best possible way.

Even if a little company has technically done the

right thing but is too small to be able to document their work effectively, that can mean trouble in the long run—especially if you're a landlord being accused of not having addressed the problem properly the first time. I provide a great deal of expert testimony in bed bug court cases, and everyone's health and legal rights are most protected by an exterminator who can provide clear documentation of their extermination work.

## Think Like Goldilocks

If you are not a corporation or a hotel, but an individual with a household, you need a pest control operator that is, as Goldilocks said about the rocking chair, not too big, not too small, but just right. They should want to educate you and make you feel more calm, not more nervous than you already are. They should deal with bed bugs regularly and be veterans at it—if you ask them how long they have been exterminating bed bugs and they tell you how long they've been in the business, ask again how long they've treated *bed bugs specifically.* Good companies tend to grow because they're good, and when they grow, they grow responsibly and without sacrificing quality control. If they've been in business for ten years and still have only four technicians, it may mean they're not getting enough word-of-mouth recommendations to sustain expansion.

When you ask them about their methods, be on the

lookout for claims that they have only one approach and it works in all situations. Whatever they use (conventional chemical, heat, freezing, etc.), no pest control operator really believes that one size fits all (all homes, all infestations, all bank accounts). The best answer they can give, as I've said, is some version of "It depends." They should have several ways of approaching it. Any responsible exterminator knows you have to plan the strategy to fit the situation, not vice versa.

## What Will This Cost Me?

A company can usually give you an estimate over the phone, assuming you're calling about your home (offices are a different story). Depending on the severity of the situation and where you live in the country, you should be spending *at least* $350 to $400 per room, and you'd be hard pressed to find a *competent* company in an urban or a dense suburban area that will do the job for less.

So be careful if it looks like a bargain—if it is, you can't afford it! A bargain often means you'll be paying more, much more, in the long run, and paying another company to do the job right. In more than 30 percent of my jobs, I am the second or third company. People have already spent thousands of dollars on companies that didn't solve their bed bug problem because they weren't incentivized to do it—but they *were* willing to come back again and again and again—for additional fees every time. (See the next section, "WWW.")

Expect that any decent service will also include the cost of a follow-up visit ten to fifteen days from the initial service. This is a necessity for dealing with residual eggs or missed bed bugs. It is possible to solve a problem in one visit if the bed bugs weren't too spread out, if the exterminator had good aim, if you caught it early enough, etc., but the odds are not with you, so you should assume that you will need at least one follow-up, and that should be part of your cost. What's the worst that could happen? The exterminator will come back and find no sign of bed bugs? We should all have such problems!

## WWW: <u>W</u>ill You <u>W</u>arranty Your <u>W</u>ork?

This is one of the most important questions you can ask. An exterminator should be willing to warranty the job for an extended period of time—ninety days is a reasonable and fair benchmark. They should promise to come back as many times as needed during that period, for no extra money.

Beware of someone who offers too *long* a warranty. No one can cast a magic spell or put a bed bug–proof force field around your home. If you get strep throat and the doctor cures you, he can't promise that you'll never get strep again.

**THE WARRANTY EXCEPTION**

The only reason not to offer a warranty is if the client is a landlord or manager of a property infested with bed bugs. With rare exceptions, the pest control operator can't know walking in where the host apartment is or what the underlying problems are. If you are a landlord or a property manager, be wary of any pest control company that says it will give you a warranty. That company cannot determine what it will take to live up to that warranty, and it may abandon you when the profit margin starts getting skinny. There are just too many variables to warranty a widespread problem.

## How Long Will All This Take?

The average length of treatment, using conventional methods, is approximately five to six weeks, from first visit to all-clear. But don't worry—your pest control operator will not be taking up residence for that whole time. After the initial treatment, any quality company will come back to re-treat your home ten to fifteen days later. If there is no more biting or other evidence (bloodstains, fecal matter, bed bug carcasses, etc.), you can bring a dog back in to inspect and clear the treated areas anytime after three to four weeks. If there *is* any evidence, you may need another treatment. Less than 15 percent of my clients have ever needed more than two treatments. It *always* gets done by three visits, because if it takes more than that, the red flag goes up—we *know* there's something going on besides what's in the space. If you find yourself needing more than three or four treatments after the first, that should be a red alert that an ongoing source of reinfestation has not been identified (boyfriend, office, houseguests, nanny) or your pest control operator is doing something wrong!

## Don't Need a Clean-Up Man!

You know those movies that feature a clean-up man, the career criminal they bring in to clean up a mess when less-experienced criminals have screwed things up? Think of Harvey Keitel in *Pulp Fiction*, or Jean Reno in *The Professional*. I hate being that guy—or the bed bug version of him. Don't get me wrong—I find it very fulfilling to help people get rid of their bed bugs (you probably figured that out already). I just hate that so many people end up paying for a service they are not even getting, *before* they come to me! Even having a "small" bed bug problem that gets caught and addressed early takes a big toll on people; how much worse to have it keep growing while you *think* it's being treated.

**WHAT EXTERMINATORS DO WRONG,
AND WHY IT HURTS YOU**

Here are some ways unqualified exterminators take your bed bug scenario from bad to terrible:

- The original exterminator(s) do not give you the right information to begin with, because they are not trained to assess your situation accurately.
- They don't tell you how to prepare your home properly.
- They use chemicals that exacerbate the problem by spreading it out.
- They use materials not even labeled for bed bug use!
- They don't look at the whole home, focusing instead on specific rooms. They need to consider all areas—in the laundry room, the basement, contiguous apartments...anyplace you go in whatever structure you live in.
- They may be treating the right areas but not thoroughly enough, because they do not have a comprehensive strategy and they are not asking the right questions. The right probing questions can help you find where your bugs came from in the first place and also let your pest control operator come up with the plan that suits your problem.
- Companies may do the right things but not in the right order. Often I'll come in, people will ask what I'm going to do. I say, "I'm going to steam

here, put some chemicals here…" They say, "No, no, that's not gonna work—that's exactly what the last guy did!" It always reminds me of the one time I tried baking a cake. I figured hey, how hard can it be? I whipped up the batter, baked it up nice, put on the frosting…Then I realized I had forgotten the sugar. I thought, *What the heck, sweet is sweet,* and put the sugar on top of the frosting. Needless to say, it didn't work out the way I planned. Many exterminators who aren't experienced with bed bugs just don't know the scope of the work, and while they may have the right weapons and the right ammo, they don't know which one to use when, and how much.

- They put down chemicals in areas they shouldn't or in improper ways, exposing people to all kinds of dangers. (Most insecticidal dust, for example, is fine inside your walls or voids, but it should not be close to where you breathe.)

I can't count the number of times I've come into homes where pest control operators had taken a bad state of affairs and made it worse. People who trusted the "professional" because they had no reason to think the work would be anything but competent. Now that you know the questions to ask and the answers to listen for, you can improve your chances of getting the help you need—the first time around!

Any decent company will tell clients that contiguous spaces have to be treated, too. It's more expensive—if you're a landlord and you find bed bugs in one apartment, you have to inspect the units around it. If you find nothing, tell tenants to start taking the preventive measures laid out in chapter 2; if you get hits, you have to treat those apartments the way you treated the first one.

## BAT BUGS

Instead of bed bugs, you may have bat bugs, which look virtually identical (you need a trained entomologist with a microscope to tell them apart). They feed on bats, though they are not too picky to feed on you if there are no bats around. Pest control operators who don't know to check will treat a home again and again and again, each time expecting better results (the classic definition of insanity). This goes back to the importance of probing questions—an experienced company, inspecting all parts of the house, will ask if you have bats or have had them recently. If they look in your attic and find bat guano (the only species I know that has a special name for its poop), they'll know what to do.

## Should I Ask "What Chemicals Do You Use?"

The short answer is "No." I get calls from people far, far away who have read about me or seen me on television. I'm already not thrilled with my two-and-a-half-hour daily commute, so Dusseldorf, Manchester, and Prague ain't happening. Sometimes my long-distance callers ask me if they should learn about all the chemical pesticides used against bed bugs so they can ask prospective exterminators more pointed questions about their methods. I always say there are much more rewarding hobbies than armchair exterminating.

Why? You don't go to dental school to have a cavity filled. You don't go to medical school so you can undergo surgery. You don't go to law school so you can have your will drawn up. (Although my mother would have been thrilled if I had gone into at least one of those professions.) You don't have to become a pest control operator to know if your bed bugs are being exterminated effectively, but if you ask the basic "What Do I Ask?" questions (see the box on pages 151–153), you'll give yourself the best odds.

Instead of turning yourself into an amateur pesticide scholar, you're better off choosing your exterminator wisely and letting him know in advance if you have any special concerns or considerations in your home, such as young children or elderly people, or anyone who has a respiratory condition, is undergoing chemotherapy, or is otherwise immunocompromised.

This will let him design a treatment that is going to be safe for your situation and also eradicate the problem.

When in doubt, go green. A green treatment—such as a combination of dry-steaming heat treatment and vacuuming—can always be used. It does not compromise the efficacy of the process, and you won't have to worry about your children or anyone else. The cost is higher (see the box "Going Green") because it is much more labor-intensive, but in all cases it is actually faster than conventional treatment, and you can rest 100 percent assured that your family is not being endangered in any way.

## Then Why Don't All Bed Bug Exterminators Go Green?

Good question. The primary reason is cost. Traditional chemical treatments, even in small jobs without a major infestation, are still expensive when you factor in all the other costs (dry cleaning, laundering, mattress covers, vacuums, lost work time, helping hands to prepare for service), which is why so many people take a stab at getting rid of this tenacious pest on their own. Green treatments typically start at fifty to one hundred percent more. Going green is great if you can afford it. Progressive and quality pest control companies have effective green programs to offer those who don't want pesticides in their home because of their lifestyle or cannot have them due to health issues.

## GOING GREEN

There *are* increasingly greener methods of exterminating bed bugs—solutions that don't rely on the chemicals most commonly used. Conventional approaches can take anywhere from four to six weeks from the first visit to clearance, depending on the intensity of the infestation. Green approaches can be faster, but they are typically two to three times more expensive because they are more labor-intensive. If you had to treat, say, two bedrooms of an apartment, the cost for conventional treatment might be in the $1,400–$1,600 range; treating the same space with green methods would cost upward of $2,400.

I personally offer an extreme green treatment using no chemicals. This treatment can eliminate bed bugs from offices or apartments within forty-eight hours. However, at the time of this writing it is necessarily more expensive because of the number of technicians and the amount of hours required to perform it.

## To Fumigate or Not to Fumigate

Even pre-PackTite, you could always treat furniture, isolate books for nine to ten months in quarantine, or fumigate everything by putting it in a truck or tenting a stand-alone house. But what is fumigation?

Basically, fumigation is sealing a space (whether your home or a truck), and depriving bed bugs of oxygen by filling the air with Vikane (sulfuryl fluoride), a colorless, odorless, tasteless gas that leaves no residue. It's safe enough to be used on food, and it breaks up into atmosphere. It's been around since 1961, and for many years has been used mainly for exterminating dry-wood termites. The advantage of structural fumigation is that it kills insects (and everything else). Everything in the chamber dies instantly with a single treatment—it works even for hoarders.

Vikane will penetrate solid objects, attacking the oxygen molecule and inhibiting oxygen-needing organisms from getting air. It's impossible for bed bugs to grow resistant, because you can't grow resistant to needing oxygen.

Fumigation is by and large a foolproof way to eliminate bed bugs in all conditions, without exception. Where heat and cryogen drive bed bugs away, and pyrethrins also need contact with bug and egg, you don't have to get the gas onto eggs, bugs, nymphs, etc., because it takes away what they need—oxygen.

If you have a freestanding home you can leave the house and have it tented and fumigated, but for an apartment or semi-attached home this is not an option. Instead, you can have your stuff fumigated off-site. You have a moving contractor do the loading, packing, and delivery, followed by canine inspection, and then fumigation of the boxed-up goods at a fumigation site. While everything that was in your home is being fumigated, a pest control operator goes into your home and treats it. That way you're bringing your bug-free belongings back into a bug-free home.

In addition there are many items that are not treatable by a technician—if you've got hundreds or thousands of books or drawers and drawers of files (you know who you are), it's not practical to PackTite all of it. Your only option would be to have it bagged up and quarantined for the next nine or ten months. But for many people—writers, lawyers, accountants— who need access to these sorts of items, the only viable option is to have them fumigated. Same thing goes for heavily infested couches and electronic devices.

But remember—while it's good that fumigating leaves no residual once the fumigant is vented, you can easily get reinfected if you bring bed bugs back from where you got them, or from where you gave them to someone else. Once you fumigate, you *have* to have a rigorous prevention program in place. (Back to chapter 2 on page 37...)

Downsides of tenting your freestanding home: it's expensive, you have to vacate for two to three days, and you may have to hire a security company to watch your home. Vikane can penetrate through wood, cloth, cardboard, etc., but not through plastic containers—if you have anything in plastic bags, open the bags first. It can only be done in temperatures above 42° so, depending on where you live, winter can be a harder time to fumigate—you have to get heat into the space for the fumigant to work properly. Some fumigators compensate for the cold by reshooting the fumigant two or three times.

Another option is as close as you can get to moving without winding up in a new home. You box up everything and put it in a truck, and the truck itself is fumigated. The pro is that you have no smell or taste of pesticide in your home. The cons are having to pack up (and ultimately unpack) everything you own, and the substantial cost. In certain situations where things are so infested that it's very difficult to treat, especially items like electronics, fumigation can be part of the best solution; but the expense makes it a luxury. Most of my clients don't opt for it, but it does make for a better job, and it expedites the whole eradication process. There's just less clutter and much easier access than when everything is in your home.

**On the way back:** Fumigating has to be done in an open lot. You can bring your own car or a trailer with your stuff to the site, but beware—if you rent a truck to pick up your stuff, the rental can itself be compromised.

For some people, fumigation is actually more cost-effective. If your dry-cleaning bill would be up in the thousands of dollars, and/or you don't want your clothes to experience the wear and tear of dry cleaning, fumigating is the way to go. Your clothes won't be any cleaner, but they will be bed bug free.

However, while you may have seen images of an

entire home wrapped in plastic for fumigating, fumigation is *not* the norm in treating bed bugs. You should not have to vacate the premises—even conventional methods should allow exterminators to treat while people are still at home. In fact, we often recommend that people spend at least two to three nights in their beds so they draw out the bed bugs onto treated surfaces. I know it means asking you to stay at home and be a human guinea pig for the greater good, but if you're not there, then bugs won't find the blood meal they are craving. Just think of it as lying in wait for them to fall into your vengeful trap—*BWAH-HA-HA-HA-HA!* (That last bit will seem crazy *only* if you haven't had bed bugs. If you have ever had them—or you have them right now—you know exactly what I mean.)

## HOT AND COLD: CRYONITE VERSUS HEAT REMEDIATION

If you haven't been keeping up with the latest in bed bug–battling technologies, this probably sounds like Mr. Freeze versus the Human Torch. Cryonite is a way of freezing bed bugs to death (think futuristic $CO_2$ fire extinguisher on steroids), and thermal remediation is a way of baking them to bed bug heaven (picture turning your home into a giant PackTite). These pages should give you an idea of what they are, why all the fuss, and why you should be suspicious if an exterminator tells you one or the other is "the only way to go!"

| CRYONITE | THERMAL REMEDIATION |
|---|---|
| If you've seen any bed bug–extermination commercials, you've heard about "miracle cure" Cryonite. Cryonite is liquid $CO_2$ in the form of dry ice snow, applied in great quantities. The technology was developed by Swedish scientists Per-Åke Hallberg and Bertil Eliasson in 1996, tested in various ways, and brought into the market by Venteco (since gone bankrupt) and Anticimex about ten years later. According to a report by one of the manufacturers, CTS Technologies AG, "it is recycled from industrial processes. Cryonite therefore does not result in additional $CO_2$ being released to the atmosphere." We're still waiting for data to back up this claim. | Thermal remediation, where a pest control company brings in high-temperature heating equipment to bake your whole home to around 135°, is becoming more popular, but it's not something you want to do as a preventive measure. |
| **PRO:** It's a great idea—exterminate insects without chemicals—and it's been marketed almost as much as the World Cup. One great feature is that you can treat electronic equipment without hurting it—an advantage over chemicals. | **PRO:** It kills all the bed bugs that aren't smart or fast enough to run to a cooler spot in your home. |

**CON:** It isn't the miracle cure you're supposed to think it is. Cryonite works only if you hit the bed bugs directly! If they're on the other side of a fabric, even a *thin* fabric like the cheesecloth on the bottom of your box spring, it won't get them—it just flushes them out of one place and into another. I've tested it myself, applying it to one side of a handkerchief with bed bugs on the other side, and the bugs did not die. Sometimes it doesn't even work with direct contact. That's a problem!

As I said in chapter 2, beware. Companies that use this technology are savvy marketers. They adopted it because it's unique, it seems green, and nobody likes the idea of pumping chemicals into their living space. It's typically far more expensive—using it to treat a studio apartment could cost from $2,500 to $3,000, a one-bedroom apartment around $4,500, and it just goes up from there. These companies have spent serious money on marketing, and they need to recoup their investment. A lot of the cost reflects what they've spent on advertising.

**CON:** It can miss spots and damage things. Vinyl, many plastics, and makeup will melt. To get the core temperature of a house to 130°, you have to raise the rest of the house to temperatures higher than that. This causes many things in the house to melt—little things you don't think about, like rubber parts in the toilet or in your furniture. Even a barely noticeable crack in a window will expand to a long ugly one, or just plain break. Bed bugs *will* scramble to pitch camp in some cold or insulated spot that the heat has not fully penetrated. To perform this service the pest control operator needs to set up large, noisy diesel generators, which will run for several hours or will make heavy electrical demands on your apartment building; many buildings cannot meet this demand.

*continued*

## HOT AND COLD: CRYONITE VERSUS HEAT REMEDIATION (*continued*)

**MY ADVICE:** If a company has Cryonite as just one tool among many in its toolbox, that's fine—it means they're trying to gauge *your* particular situation and work out the best combination of tools for solving it. But I would not recommend any exterminator who uses it exclusively, or even primarily—if you feel an exterminator is pushing Cryonite on you ("We've never had a situation where Cryonite didn't work..."), find another one. It's not a stand-alone product. I don't oppose the use of Cryonite in specific situations, but I have found dry steam superior in a variety of ways. It's always green, there's no environmental question mark, it's easy to transport, and the data has proven that it kills bed bugs effectively.

**MY ADVICE:** It's a lot of money to spend on something that ultimately can't guarantee total extermination of the pest.

Both technologies are highly controversial, and the minority of the industry has adopted them, due to their unreliable effectiveness and the numerous problems that can cause them to fail. Based on the facts and my overall experience, in most situations I have to give them two thumbs down.

Cryonite is ultimately no more effective than a spray bottle of Steri-Fab. You use it by spraying it directly on bugs. Thermal remediation will kill a lot of bugs as well, but it will also disperse them. With both these treatment methods, you see nothing for a while, but then the bugs come crawling back. If you get treated again, you just have a roller-coaster effect. Don't you have enough troubles already?

I'm *always* striving to be on the cutting edge of environmentally friendly methods for eradicating bed bugs (it's one of the main reasons I got into the business, subsequently stunning my mother when she realized there was no law school in my future), so I totally understand the potential appeal of Cryonite and heat remediation. Both approaches have their good sides, but they are somewhat costly and beyond many people's budgets. Even worse, neither has any residual power (unlike conventional pesticides); if a bed bug escaped the treatment, nothing in the aftereffect will kill it. You'd be forced to bring back the whole apparatus again—and you'd have to start all over! Conceptually thermal remediation is a great idea, but it's far from perfect and it's not yet at 100% efficacy. Perhaps time and research will perfect this technology, but until then you're wise to think carefully before choosing it. I've seen too many clients disappointed and lighter in the wallet when their bed bugs were treated with either of these methods.

In your search for the pest control operator who can eradicate bed bugs from your space, just remember: the situation should determine the technology, and not the other way around. Your best exterminator is not one who has the heaviest and shiniest artillery, but the one who is experienced, dogged, fair, and flexible. The right exterminator for you will tailor the treatment to your needs and your means; he will help you help yourself; and he will keep coming back until the job is done. Anything less is not worth your time or your money.

# PART THREE

# The Past and Future

Where Our Buggy Friends Have Been, and Where They Are Going (Other than Hell, of Course)

# 8

# A Brief, Illustrated History of the Bed Bug

## How It Has Plagued Us and How We Have Tried to Get Rid of It

You've made it this far. And when it comes to bed bugs, you know a whole lot about how to avoid them, save money if you have them, and generally, not freak out. But now that you've gotten on such a chummy basis with this little crawler, aren't you just a wee bit curious about where they were before they invaded the local cineplex, office building, or residence?

Bed bugs, which plagued man since antiquity, are technically called *Cimex lectularius*, Latin for "bug of the bed." They also go by the names wall louse, mahogany flat, crimson rambler, heavy dragoon, chinche, redcoat, and, if you listen to Kenneth the Page on the NBC sitcom *30 Rock*, chew daddies and Mugabe's concubines. They are believed to come from the Middle East, living in caves inhabited by bats and humans. They once had wings that they never need to use because their food source—man—was right there alongside them.

And here is where they've been ever since:

**1352 BC:** The earliest fossil evidence of bed bugs comes from a 3,550-year-old archaeological site in Amarna, Egypt, believed to be the home of the workers who built the pyramids.

**423 BC:** "I have a bum-bailiff in the bedclothes biting me." Followed later by the lines "Oh! I am a dead man! Here are these cursed Corinthians advancing upon me from all corners of the couch; they are biting me, they are gnawing at my sides, they are drinking all my blood, they are yanking on my balls, they are digging into my arse, they are killing me!" (Aristophanes, *The Clouds*)

**405 BC:** Dionysus, asking advice from Heracles "about ports, towns, brothels, bakeries, restrooms, roads, where to get a drink, landladies, and lodgings with the fewest creepy crawlies." (Aristophanes, *Frogs*)

**AD second century:** "I tell you, you bugs, to behave yourselves, one and all; you must leave your home for tonight and be quiet in one place and keep your distance from the servants of God!"—*Acts of John* (New Testament apocrypha)

**AD 77:** In the exhaustive encyclopedia of the ancient world, *Natural History*, Roman officer and author Pliny the Elder promotes the medicinal uses of bed bugs, including giving seven bed bugs in a cup of water to adults suffering from lethargy. According to Pliny, they also help heal snakebites and warts.

**1000s:** Bed bugs first mentioned in writing in Germany.

**1200s:** Bed bugs hit France.

**Circa 1600:** At a time when *bugs* were synonymous with *bed bugs*, Shakespeare writes "Such bugs and goblins in my life" in *Hamlet*.

**1666:** Rumors abound that bed bugs took over London after being transported in with supplies of wood used to rebuild the city after the Great Fire of London.

**1700s:** Colonists bring bed bugs over to the not-so–New World by ship, leading to mass infestations in new settlements.

**1730:** A British book, *A Treatise on Buggs*, includes a line "Reasons given why all attempts hitherto made for their destruction have proved ineffective." Great.

**1851:** A pest management company in the United Kingdom nicknames itself "Bug destroyers to her Majesty and the Royal Family" because, yes, the royals had bed bugs, too.

**1927:** "Gals, bed bugs sure is evil, they don't mean me no good / Yeah, bed bugs sure is evil, they don't mean me no good / Thinks he's a woodpecker and I'm a chunk of wood." (Bessie Smith, "Mean Old Bed Bug Blues")

**1930s:** Distressed Southerners in the United States tried to fight bed bugs by placing their bedposts in small cans filled with kerosene—and occasionally setting them on fire to kill off any stubborn bugs.

**1933:** A report by the UK Ministry of Health claims that in some neighborhoods all the houses had some degree of bedbug infestation.

**1935:** "The bedbugs bothered him a little at first, but as they got used to the taste of him and he grew accustomed to their bites, they got along peacefully. He started playing a satiric game. He caught a bedbug, squashed it against the wall, drew a circle around it with a pencil and named it "Mayor Clough." Then he caught others and named them after the City Council. In a little while he had one wall decorated with squashed bedbugs, each named for a local dignitary. He drew ears and tails on them, gave them big noses and mustaches." (John Steinbeck, *Tortilla Flats*)

**World War II:** Not surprisingly, bed bugs are not a welcome part of the war effort during World War II. General Douglas MacArthur says that they were the "greatest nuisance insect problem...at bases in the U.S."

**1940s:** After World War II, DDT, dusted on or sprayed around beds, becomes a more effective, though not foolproof, deterrent to bed bugs.

**1950s:** DDT gets an upgrade when an organophosphate insecticide is added to it. It proves highly effective against the bugs, and eager Westerners even begin adding DDT behind the wallpaper in their nurseries.

**1972:** DDT is banned. But bed bugs don't even notice—

by that time, they had already been nearly wiped out in the United States for twenty years.

**1996:** A Manhattan hostel calls me, curious to know what was causing the complaints of itching from all of their clients. And that's how I encounter my first bed bugs and realize that this once-eradicated pest is back. Around this same time, others in the pest control industry are noticing something is up.

**2005:** The *New York Times* runs a front-page story, called "Just Try to Sleep Tight, The Bed Bugs Are Back," about the bugs that are spreading through the city "like a swarm of locusts." The rest of the nation takes note and thanks its lucky stars that they don't live in the Big Apple, never imagining it could spread to their areas.

**2006:** Bed bugs become a new and popular big-deal topic at pest control conventions.

**2006:** In an episode of the sitcom *The King of Queens* called "Buggie Nights," an exterminator hired to rid the couple of their bed bug infestation says, "Sleep tight...I think you know the rest!"

**2007:** New York lawyer Sidney Bluming, and his wife, Cynthia, sue the Mandarin Oriental Hotel Group after getting bitten hundreds of times during a five-day stay at a location in London—and taking the bugs home with them across the pond. The Blumings, who are seeking several millions of dollars, release a statement that says, "People associate bed bugs with more of a lower-end class of hotel. Clearly that's not the case here..."

**2008:** Bed bugs head to college, infesting dorms from New York University to North Carolina's Guilford College and others from Kansas and Missouri.

**2009:** The *Daily News* reports that President Clinton's offices in Harlem are infested with bed bugs. My company and I couldn't be happier to help out the former president when members of his staff call us in.

**2009:** Entomologist Dr. Shripat Kamble of the University of Nebraska tells a bed bug convention that he remembers the following remedies being used during his childhood in India: "People commonly used in the summertime heat treatment. Keeping the cot outside in the hot sun," and shaking the bed so the bugs spilled onto bare ground hot enough to kill. Also popular was to pour "boiling water through all the hiding areas of the bed bugs...A lot of times it worked, and sometimes we still had problems."

**2009:** "I don't have bed bugs...I went to Princeton," laments an in-denial Jack Donaghy, the NBC exec played by Alec Baldwin on *30 Rock*. In this episode he battles a bed bug problem and faces the social ostracism of his fellow New Yorkers.

**2010:** Terminix releases its survey of the top ten most bed bug–infested cities, showing that it is not just a New York problem but plagues all American cities, including Philadelphia, Detroit, Cincinnati, and Chicago.

**2010:** The Morristown, New Jersey, Housing Authority announces that it will begin doing public housing inspections every three months. Marion Sally, executive director of the Housing Authority, describes the policy as "a preventive measure" made necessary because too often by the time residents "call, it's already become a serious problem. Sometimes they don't call because they feel ashamed or they may think there are bed bugs because of dirty conditions."

## DON'T LET THE BED BUGS BITE: CURIOUS ABOUT THE CHILDREN'S RHYME?

*Good night, sleep tight,*
*Don't let the bed bugs bite.*
*And if they do*
*Then take your shoe*
*And knock 'em 'til*
*They're black and blue!*

If you've never sang this nursery rhyme to your own kids, your parent likely sang it to you. Like many childhood ditties, it is gruesome and gross. However, it is also chock-full of historical info.

The phrase *sleep tight* explains how bedding worked until the 1940s, when DDT gave people another way to fight bed bugs. *Sleep tight* referred to the rope beds people slept on. Beds were often square frames elevated from the ground, with ropes tied across in a weave pattern, almost a hammock. In order to sleep well, the "mattress" couldn't sag, so the ropes had to be periodically tightened to keep the bed "tight."

**YOU'RE NOT THE ONLY ONE WHO HATES THEM:
THE BED BUGS' NATURAL ENEMIES**

Although it may feel like it, bed bugs are actually not invincible. Here's what's out to get them, at every turn:

- cockroaches
- mice
- ants (especially the Argentine, Pharaoh, Tropical fire, and European imported fire)
- certain spiders
- mites
- centipedes

In the hopes of garnering a few other enemies of the bed bug, in the 1800s, mattresses, according to Charles Panati in *Extraordinary Origins of Everyday Things*, were made of organic materials such as "straw, leaves, pine needles, and reeds" and tended to rot, mildew, and harbor rats and mice, who were hunting for bugs!" (By the 1870s, spring mattresses came into popular use.)

# 9

## The Future and *You*

What We Can All Do to Stop Bed Bugs
in Their Tracks

If you're like most people, the first thing that came to
mind when you heard bed bugs were making a creepy-
crawly resurgence was, *What do I need to do so that
I don't get them?!* And it only makes sense that everyone thinks of protecting themselves without thinking
about others—after all, you're on your own when it
comes to scratching your bites and paying for extermination if you find them in your bed. But I'd like to
think that by now, you understand that this is bigger
than any one person. Now that you know how to save
*yourself*, it's time to think about the greater good.

In this chapter, I'll tell you what's on the horizon
for bed bug eradication—from grassroots movements
to fledging policies to new detection techniques—and
how you can help the cause. Because this is a community problem that will take *everyone's* help to solve.

## CHECKLIST: The Six Things *You* Can Do to Save the World from Bed Bugs

Here's where "I" meets "we." If you do these things and nothing else, you will be actively working to stop (or at least slow) the spread of bed bugs for everyone's sake:

- **Bring in dogs periodically.** This will keep you from unwittingly passing on an undetected infestation to visitors.
- **Don't ignore bugs, or play "Wait and See" roulette.** I've seen thousands of potential clients make that gamble, and without fail they came crawling back with an infestation that was much tougher (and more expensive) to treat than if they'd just let me (or another qualified pest control operator) treat it in the early stages.
- **Take advantage of the products that work.** I wrote earlier about how our culture is obsessed with shopping. Well, this is one time when that obsession could help us all. Buy (and use) the repellents and pesticides (like JT Eaton's Rest Easy and Steri-Fab) to prevent an infestation in your space that others could carry home.
- **Travel safely.** Doing hotel room inspections, properly sealing your luggage, and avoiding the places bed bugs like to hang out will all slow the spread of bed bugs. You definitely don't want to be the per-

son who introduced them to your seatmates on the flight home from your favorite vacation spot.

- **Be a good host.** Fib about the reason if you have to, but if you even *suspect* you have a problem, don't have guests in your home. Period. A little white lie is infinitely more ethical than having them sleep over and your saying nothing.
- **Exercise caution.** If you have bed bugs, be incredibly careful not to take them with you, whether you're going to work or to the movie theater. Heat and bag any clothing or other items before you take them out of your home, so you can avoid spreading them to your office, where an unsuspecting coworker could carry them home.

## Advocate, Advocate, Advocate

While it's clear that the government should be doing more at all levels, what isn't always as clear is exactly *what* they should be doing. I spoke with Washington, D.C.–based policy consultant Larry Pinto and Bob Rosenberg, the senior vice president of government affairs at the National Pest Management Association (NPMA) about just that, and the one word that kept coming up was "education," which, coincidentally, is my sole purpose for writing this book. It is truly the first step when it comes to preventing and treating infestations. An effective program must not only include public education but also involve working with industry groups (including hoteliers, transportation organizations, and used furniture purveyors) to help them understand the urgency involved, and fostering good relationships with qualified pest control companies so they can develop industry-specific best practices. There also needs to be proactive education programs targeted to folks that deal with transient populations, including local social service agencies (such as shelters, foster boarding homes, and mental health facilities), universities (dorms, ick), nonprofits, hospitals (can doctors *please* get training in identifying and treating bites, and stop telling patients to bomb their homes?!) and senior care centers. Best case, they would come out of training with the tools they need to put together both a prevention strategy and a plan of attack in case of infestation. And, of course, these programs need to be well funded to be effective.

Asking the government to develop regulations or best practices isn't ideal, because government agencies are notoriously bureaucratic, and by the time they come up with recommendations, new methods will likely have been developed—they will usually be behind the curve, and the people will suffer for it. But we *should* ask our congressional representatives to push for consumer protection when it comes to all those so-called bed bug treatments on the market. So many of them are either dangerous to your health or send bugs into hiding, where they bide their time until it's all clear to come out and feed; they need to be pulled from the market, or at least be accurately labeled so customers know what they're really getting.

It's also key that you know what you're getting when you choose a pest management company. As of this writing, the NPMA was convening a Blue Ribbon Bed Bug Industry Task Force to study the pros and cons of developing industry-policed training or certification for operators who treat bed bugs, so it's possible that infestation victims will soon be able to use certification as a measuring stick when choosing someone to treat an infestation.

But much of the practical work and regulation enforcement will likely fall to local agencies as municipalities are being overrun with reports of infestations. Cincinnati, Toronto, Baltimore, and other cities have created action plans tailored to their residents, and all major urban centers (at the very least) will need to follow suit if we're going to put my bed bug practice out of business.

## Go Grassroots

You can join the anti–bed bug movement right from your computer. A few cities have privately run, nonprofit advocacy groups; find one in your area by searching online for "bed bug advocacy" in your city or state.

No advocacy group in your area? Organize! It worked for New York vs. Bed Bugs, the advocacy group that a handful of city bed bug sufferers formed in early 2008. They disbanded two years later after they achieved their goal of pushing the city to create a plan for fighting bed bugs, but their website (NewYorkVsBedBugs.org) is still a great resource for people looking to start their own groups. So is the Central Ohio Bed Bug Task Force's site, Central OhioBedBugs.org. Not only can you push the government to take action, but you can provide solace for stigmatized bed bug sufferers who are silently scratching in the shadows.

Are bed bugs hopping from apartment to apartment in your building, like they're playing some deranged game of checkers? Start a tenant organization so that you can work together to get your landlord to treat the property systematically for everyone's sake. Just pick a meeting time and place (preferably outside your building) and go door to door with flyers explaining what you're trying to do. At the very least, you'll help your neighbors understand how big an issue this is—or will be—for them. Then be tough and persistent. I have seen countless tenant groups who organized to get a landlord to take proper action, and the landlord was less than accommodating, even calling tenants dirty and denying any responsibility. But some landlords just needed to be nudged to find a competent exterminating company to not only treat the problem systematically but also develop a comprehensive strategy to prevent reinfestation. Don't be afraid to push.

If bed bugs are a scourge on your community and your local government seems remiss to do anything about it, you can join with other concerned citizens and interested parties (namely, people who have fought bed bugs in their own homes, city councilmen up for reelection, local pest control companies, and university entomologists) and attempt to force their hand. Create a website for the cause, then reach out to people who (should) care. Be sure to include a way for people to report infestations (so you can compile statistics to use in your fight) and links to resources, such as action plans being put in place in other cities and the latest articles, studies, and research. Join the USvBB mailing list to connect with other organizers who are working for the cause. You can find details in the resources section on NewYorkVsBedBugs.org.

Next, contact your local health and housing agencies and ask if they are keeping a record of infestations. If they aren't, insist that they establish a way to take reports from private citizens who encounter bed bugs in their rented homes, in their places of employment, and in other places of business. The media seems to love stories of bed bug infestations; take advantage of their hunger for a good story and direct reporters to your site so they can fight the battle in the public forum—the more people who know what's going on, the more allies you'll be able to enlist. Good luck.

On an individual level, don't be afraid to ask the people with whom you spend money what they are doing to prevent bed bugs. Ask your therapist if she's had to treat her couch, find out if your primary care doc has a prevention program in place, and inquire whether your favorite consignment store inspects all pieces before putting them on the sales floor. Service providers will not care about this unless their clients and customers show that *they* care. Who knows, maybe one day we'll have the equivalent of a Good Housekeeping Seal of Approval for businesses with proactive bed bug prevention programs.

## On the Horizon

If you've learned nothing else from reading this book, you've learned that there is no magic bed bug–killing silver bullet that can eradicate these pests forevermore. But there are some great products that can help you easily integrate prevention into your lifestyle and help keep you pest free. Even as these new products are coming down the pipeline, the pros are *really* hoping that research at both the university and the corporate levels will help us create effective infestation alarm systems (so you'll know you have a problem before you're bitten) and more efficient ways to treat for them once you discover an infestation.

The Holy Grail would be the equivalent of roach motels for bed bugs. The best pest control technology in the past thirty years was the development of baiting technology, which was first used to kill roaches—you know it as Combat. It contained a very low level of toxicant that killed roaches that fed on it (and they, in turn, took the egg producers with them). Its immense success led to specialized baits for ants, carpenter ants, termites, and flies, and they are a green method of extermination. If the entomological science world could develop a bait for bed bugs, we'd be much closer to controlling the bed bug epidemic.

The problem? While you can lure bed bugs in with carbon dioxide, there's (currently) no way to keep poisoned blood at precisely the right conditions (namely warm, fresh, and inside a human) to entice them to feed. I spoke with a dream team of entomologists—Dr. Changlu Wang at Rutgers University, Dr. Phil Koehler at the University of Florida, and Dr. Josh Benoit at Yale—and while it's clear the field is advancing rapidly, we're no closer to developing a viable bait now than we were ten years ago.

When it comes to treatments of the future, Dr. Koehler has had some success with building heat boxes around beds that can contain and kill bed bugs within twenty-four hours. But as of this writing, the process was too labor-intensive and costly to be replicated for the masses. Dr. Benoit has been researching how bed bugs survive between meals. He believes that if entomologists can decipher a way to reduce the time bed bugs can live without biting, we can limit the bug's ability to spread and thrive.

In the realm of detection, Dr. Wang's research has shown that homemade carbon dioxide detectors are extremely effective in the laboratory—though I find they are somewhat less effective in the field. They can be created for less than twenty dollars and are actually better at attracting bed bugs than expensive detectors on the market that combine $CO_2$ with heat and chemicals. But when concerned consumers build them by hand, they can be dangerous, and even small mistakes can render them virtually useless. It's almost a no-brainer that some enterprising company will eventually develop a kit that standardizes the process.

## *The Bed Bug Survival Guide* and You

Use the BedBugSurvivalGuide.com website as an instructive resource. It's an educational, informational clearinghouse for all things bed bug related. There will be articles, blogs, and up-to-the-minute resources in the war against bed bugs. All proceeds from this website are donated to charity.

## The Bottom Line

Battling bed bugs can feel like the loneliest, most isolating task ever. But if we band together, we can eradicate this pest (and the stigma that it carries). To review:

- You can help thwart the bed bug takeover by doing six totally unselfish things (pages 186–187).

- All levels of the government will need to be involved to solve this problem (pages 187–189).
- You can join or start an anti–bed bug movement and be part of the solution (pages 189–192).
- If we're lucky, today's research will create the bed bug killer tech of tomorrow (pages 192–194).

# Make a Federal Case of It

Wonder what the federal government is doing to kill bed bugs dead? The short answer is that there are a lot of plans, but it remains to be seen how they will be put into action (and how well funded said action will be). The Bed Bug Federal Working Group, an interagency force aimed at addressing the bed bug issue, was established in 2009; the Environmental Protection Agency, the Centers for Disease Control and Prevention, the Department of Housing and Urban Development, the National Institutes of Health, the Department of Agriculture, and the Department of Defense are all involved. The group was formed as a result of recommendations that came out of the EPA's 2009 National Bed Bug Summit (see the table on the following page), and it aims to work with state, tribal, and local health departments, academia, and private industry to monitor and research bed bugs in the United States.

Here's what individual agencies and branches of government are doing.

| AGENCY/ARM OF FEDERAL GOVERNMENT | WHAT THEY ARE DOING—OR PLAN TO DO—ABOUT BED BUGS |
| --- | --- |
| **United States Congress** | • Representative George Butterfield (D-NC) introduced the Don't Let the Bed Bugs Bite Act of 2009 in May 2009 (originally introduced as the weaker Don't Let the Bed Bugs Bite Act of 2008; now dead, obviously). It called for the establishment and funding ($50 million a year from 2010 through 2013) of a grant program, administered by the Secretary of Commerce, to help states train bed bug inspectors, inspect hotel rooms for bed bugs, contract pest control operators, and educate hoteliers on how to prevent and treat infestations. It would have also added bed bug prevention and management to public housing agency plans, and required the CDC to research and report on the public health effects of bed bugs. It was referred to the House Financial Services and House Energy and Commerce committees, where it was languishing as of this printing. |
| | • Senator Charles Schumer (D-NY) wrote to the heads of the EPA, HUD, and HHS, asking them to form an interagency task force to "investigate and resolve" bed bug infestations across the country. It was formed, if unofficially. |
| | • Congresswoman Jean Schmidt (R-OH) introduced the To Assist the State of Ohio in Conducting a Bed Bug Prevention and Mitigation Program bill (great title, right?) in late 2010 just before Congress recessed. If successful, it will establish a pilot program (funded by an EPA grant) that would help housing authorities in Ohio address persistent infestations in low-income areas. |

| | |
|---|---|
| | • Congressmen Butterfield and Don Young (R-AK) hosted a Congressional Bed Bug Forum in late 2010 that examined what the government can do to address the problem, including identifying grant programs that can fund the creation of anti–bed bug programs; examining the hurdles that block new, efficient control products from the market; investigating the risks associated with consumer products that are sold with bed bug–killing claims; and discussing the merits of requiring universities that receive federal funding to research bed bug–control methods. |
| **Environmental Protection Agency (EPA)** | • Coleading the interagency charge against bed bugs. |
| | • Issued a joint statement with the CDC in August 2010 that stated that bed bugs are "a pest of significant public health importance" that "cause a variety of negative physical health, mental health and economic consequences." This is important, because part of the reason infestations have been able to flourish is that the government has seen bed bugs as more of a nuisance than a threat to health. |
| | • Ensuring that pesticides used to treat bed bugs are both safe and effective. |
| | • Working with pest management pros and the public to disseminate the latest info on tools that can be used to fight bed bugs. |
| | • Working with the pest control industry and researchers to identify new pesticides to kill bed bugs. |

*continued*

| | |
|---|---|
| | • Established the "State and Tribal Assistant Grants—Bed Bug Education/Outreach and Environment Justice" in September 2010. The $550,000 total allocation went to projects aimed to build education and outreach programs "to reduce infestations in communities disproportionately exposed to environmental harms and risks." The programs that are awarded grants will be made available for replication in other similarly challenged communities. |
| | • Held National Bed Bug Summits in 2009 and 2010, which brought together pest management pros, universities, federal, state, and local government agencies, and public health organizations to develop recommendations for addressing major bed bug issues. |
| **Centers for Disease Control and Prevention (CDC)** | • Coleading the interagency charge against bed bugs. |
| | • Issued a joint statement with the EPA on the government's role in treating bed bugs (see the EPA section of this table). |
| | • Partnering with experts in health, entomology, epidemiology, and environmental toxicology fields to research how we can better control bed bugs. |
| | • Reporting info on trends in treating bed bugs with an eye toward developing a national strategy to reduce infestations nationwide. |
| | • Encouraging increased bed bug research to determine causes for resurgence, the potential for bed bug–disease transmission, and effective methods of eradicating them. |

| | |
|---|---|
| **Department of Housing and Urban Development (HUD)** | • Funding bed bug monitoring and control research for low-income, multifamily housing.<br><br>• Educating public housing authorities on bed bug identification and treatment. |
| **Department of Agriculture (USDA)** | • Supporting the Bed Bug Federal Working Group with statistical data and entomological research from staff scientists, specifically as it relates to establishing occurrence and prevalence, and measuring impact throughout the country.<br><br>• Helping to develop pesticides that could be effective in controlling bed bug populations. |
| **National Institutes of Health (NIH)** | • Providing the Bed Bug Federal Working Group with input regarding current and future potential bed bug research, and aiding in the organization of scientific meetings. |
| **Department of Defense (DOD)** | • Providing support from an army entomologist who contributes technical expertise on both the biology and control of bed bugs. |
| **Department of Transportation (DOT)** | • Nothing directly, but their "air curtain" technology could potentially be used to stop bed bugs from entering plane cabins. (They created this technology while looking for alternatives to pesticides for use on international flights going to countries that require the dangerous practice of on-board pest treatment before landing. These countries would use a blast of air at passenger entrances, to blow off flying insects hitchhiking on clothing, and a net to trap them at service entrances.) It was still in development and testing phase as of this writing, but there were no plans to test it for bed bug–prevention efficacy. |

*continued*

| **Federal Aviation Administration (FAA)** | • Nothing. There are no rules in the United States that force airlines to limit bugs (bed or otherwise) in the passenger area on airplanes. Furthermore, when I spoke to them while I was writing this book, they were not (yet) worried about bed bugs on airplanes. I wonder how long they'll be able to say *that*. |
| --- | --- |

# About the Author

With almost two decades of experience and expertise, JEFF EISENBERG, the founder of New York City–based Pest Away, is one of the leading bed bug experts and exterminators in the country. Eisenberg and his company have been featured in the *Sunday Times Magazine*, and on CNN, *The Today Show, Dateline*, the CBS *Early Show*, MSNBC, ABC News, and Fox News. Pest Away was selected from among thousands of pest control companies as the "BEST IN NY" by *New York* magazine. He has written bed bug protocols for dozens of corporations and federal government agencies, and he is regularly asked to serve as a consultant or expert witness all across the U.S. on the bed bug epidemic.

www.pestawayinc.com